Fluid Flow, Heat and Mass Transfer at Bodies of Different Shapes

T0335570

Fluid Flow, Heat and Mass Transfer at Bodies of Different Shapes

Numerical Solutions

Kuppalapalle Vajravelu
The University of Central Florida, USA

and

Swati Mukhopadhyay
The University of Burdwan, India

AMSTERDAM • BOSTON • HEIDELBERG • LONDON
NEW YORK • OXFORD • PARIS • SAN DIEGO
SAN FRANCISCO • SINGAPORE • SYDNEY • TOKYO
Academic Press is an imprint of Elsevier

Academic Press is an imprint of Elsevier
125 London Wall, London, EC2Y 5AS, UK
525 B Street, Suite 1800, San Diego, CA 92101–4495, USA
225 Wyman Street, Waltham, MA 02451, USA
The Boulevard, Langford Lane, Kidlington, Oxford OX5 1GB, UK

Notices
Knowledge and best practice in this field are constantly changing. As new research and experience broaden our understanding, changes in research methods, professional practices, or medical treatment may become necessary.

Practitioners and researchers must always rely on their own experience and knowledge in evaluating and using any information, methods, compounds, or experiments described herein. In using such information or methods they should be mindful of their own safety and the safety of others, including parties for whom they have a professional responsibility.

To the fullest extent of the law, neither the Publisher nor the authors, contributors, or editors, assume any liability for any injury and/or damage to persons or property as a matter of products liability, negligence or otherwise, or from any use or operation of any methods, products, instructions, or ideas contained in the material herein.

ISBN: 978-0-12-803733-1

British Library Cataloguing in Publication Data
A catalogue record for this book is available from the British Library

Library of Congress Control Number: 2015939630

For information on all Academic Press publications
visit our website at http://store.elsevier.com/

Contents

Part II   Further Applications   121

Preface

Fluid mechanics is one of the oldest branches of applied mathematics. It is also the foundation of the understanding of various aspects of science and engineering. A wide variety of mathematical problems, appearing in areas as diverse as fluid mechanics, mechanical engineering, chemical engineering, theoretical physics, and aerospace engineering, have been solved by means of analytical or numerical methods. Though analysis of different types of fluid flow and heat or mass transfer problems are available in the open literature, there are still a number of gaps that are to be filled up. There are several excellent books covering different aspects of fluid flow and heat or mass transfer. Yet one still looks for a systematic and sequential analysis that helps in understanding this particular area of interest.

To help students and researchers acquire a deeper understanding of the characteristics of fluid flow and heat and mass transfer, this monograph aims to present, in general, a study of transport phenomena. It is well known that for external flows, the shape of the object influences the flow over an object (i.e., a body) significantly. As a result, it affects the heat and mass transfer characteristics. In other words, the book aims to help readers develop their understanding in this particular field without spending huge time in searching the endless literature on this area. To help develop a clear insight, we discuss several flow features. By maintaining the applicability of the obtained results, we also discuss several cases of physical problems.

In selecting specific problems to work through, we have restricted our attention to the phenomena of fluid flow and heat or mass transfer as such problems introduce a wide variety of mathematical problems of interest. Hence, in order to illustrate various properties and tools useful in analyzing the problems, we have selected recent research in the area of fluid flow and heat or mass transfer.

We appreciate the support and motivation of the editor Glyn Jones and the editorial project manager Steven Mathews. We also acknowledge the role of Elsevier (Oxford) for making this book a reality. Thanks to Mr. Sudipta Ghosh (PhD student of Dr. Swati Mukhopadhyay) for his help in drawing some of the figures. We thank Prof. Mike Taylor for reading the entire manuscript and suggesting some needed changes. The authors are grateful to all the authors of the articles listed in the bibliography of this book. The authors are also very much thankful to their coauthors. Finally, we very much like to acknowledge the encouragement, patience, and support provided by the members of our families.

K. Vajravelu
Orlando, Florida
S. Mukhopadhyay
Burdwan, India
2015

Introduction

Air and water are the most important constituents of the environment we live in, so that almost everything we do is connected to the science of fluid mechanics. For example, the flight of birds in the air and the motion of fish in the water can be explained from the perspective of fluid mechanics [1]. The designs of airplanes and ships are based on the theory of fluid mechanics. Fluid mechanics is one of the oldest branches of applied mathematics, and the foundation of the understanding of different aspects of science and engineering [2–5]. From the nineteenth century, the scope of fluid mechanics has steadily broadened, as the study of hydraulics was associated with the growth of the fields of civil engineering and naval architecture. In recent times, the development of the different branches of engineering, namely, aeronautical, chemical, and mechanical engineering, have given additional stimuli to the study of fluid mechanics. It now ranks as one of the most important basic subjects not only in applied mathematics but also in engineering [6]. Now, it is a subject of widespread interest in almost all fields of engineering as well as in astrophysics, meteorology, physical chemistry, plasma physics, geophysics, biology, and biomedicine [7–9].

In nature, fluid flow over bodies occurs frequently and gives rise to numerous physical phenomena, for example, drag force acting on trees, underwater pipelines, automobiles, the lift generated by airplane wings, upward draft of rains, dust particles in high winds, and transportation of red blood cells in blood flow (see, [6]). Sometimes, fluid moves over a stationary body, for example, wind blowing over a building, or a body moving through a quiescent fluid or a bus moving through air. Such motions are referred to as flows over bodies or external flows [10]. The shape of the object has profound influence on the flow over a body and thus affects significantly the heat and mass transfer characteristics. Flow past bodies can be classified into incompressible and compressible flows. Compressibility effects are neglected at velocities below 360 km/hour, and such types of flows are known as incompressible flows. In this book, we are concerned with incompressible fluid flows [11–17].

Because of the recent high demand in the need for understanding and analyzing the problems we come across in science and engineering, we feel that there is a need for a book of this kind. The underlying aim of this book is to present transport phenomena that will help students and researchers in the field of fluid mechanics in acquiring a deeper understanding of the characteristics of flow and heat and mass transfer (see, e.g., [18]). Obviously, part of the material in the book can be conveniently used as an introductory course material for researchers working in boundary layer theory, flow, and heat and mass transfer [19–45]. Also, the book is intended for graduate students in mathematics, engineering, and in the mathematical sciences. In addition, the material in the book may be of interest to researchers working in physical chemistry, soil physics, meteorology, and nanotechnology.

The book is designed to accommodate several topics of varying emphasis, and the chapters comprise fairly self-contained material from which one can make various coherent selections. There are several underlying themes that become apparent when one examines the literature on the subject. We hope to bring out clear insight by discussing several flow features. Also, we discuss several cases of physical problems in general.

The outline of the book will be as follows. Chapter 1 deals with the numerical method(s) adopted in these works. In Chapters 2–4, which comprise Part I of the book, we present the flow past surfaces of different types, namely, stretching, shrinking, and flat surfaces. This first set of chapters provides explanations intended for general readers and can be directly employed for problems in engineering, applied physics, and other applied sciences. We keep the discussion broad based so as to provide a framework for researchers. In order to motivate the reader and provide a good understanding of the subject matter, at the end of Chapters 2–4, we provide multiple examples of problems that have been solved numerically.

In Part II of the book, Chapters 5–6, we shift the focus to concrete examples and problems related to bluff bodies in fluid mechanics and heat and mass transfer. Here the governing equations of the problems are highly nonlinear differential equations. The problems considered in this Part will help the reader understand problems of physical relevance and apply it to the other physical fields of interest. We group such problems into three chapters: general fluid flow past a cylinder in Chapter 5, fluid flow over a sphere in Chapter 6, and finally problems related to flow past a wedge in Chapter 7.

References

[1] Yuan SW. Foundations of fluid mechanics. New Delhi: Prentice-Hall of India; 1969.
[2] Abbas Z, Wang Y, Hayat T, Oberlack M. Slip effects and heat transfer analysis in a viscous fluid over an oscillatory stretching surface. Int J Numer Meth Fluid 2009;59: 443–58.
[3] Andersson HI, Bech KH, Dandapat BS. Magnetohydrodynamic flow of a power law fluid over a stretching sheet. Int J Non-Linear Mech 1992;27:929–36.
[4] Ariel PD, Hayat T, Asghar S. The flow of an elastico-viscous fluid past a stretching sheet with partial slip. Acta Mechanica 2006;187:29–35.
[5] Bachok N, Ishak A, Pop I. Mixed convection boundary layer flow over a permeable vertical flat plate embedded in an anisotropic porous medium. Math Problems Eng 2010;659023:http://dx.doi.org/10.1155/2010/659023.
[6] Batchelor GK. An introduction to fluid dynamics. London: Cambridge University Press; 2006.
[7] Bhattacharyya K, Mukhopadhyay S, Layek GC. Slip effects on an unsteady boundary layer stagnation-point flow and heat transfer towards a stretching sheet. Chin Phys Lett 2011;28(9):094702.
[8] Bhattacharyya K, Mukhopadhyay S, Layek GC. MHD boundary layer slip flow and heat transfer over a flat plate. Chin Phys Lett 2011;28(2):024701.
[9] Bhattacharyya K, Vajravelu K. Stagnation-point flow and heat transfer over an exponentially shrinking sheet. Comm Nonlinear Sci Numer Simulat 2012;17(7):2728–34.

[10] Cebeci T, Bradshow P. Momentum transfer in boundary layer. Washington, DC: McGraw-Hill Hemisphere; 1977.

[11] Chamkha AJ, Aly AM, Mansour MA. Similarity solution for unsteady heat and mass transfer from a stretching surface embedded in a porous medium with suction/injection and chemical reaction effects. Chem Eng Comm 2010;197:846–58.

[12] Cortell R. Flow and heat transfer of a fluid through a porous medium over a stretching surface with internal heat generation/absorption and suction/blowing. Fluid Dyn Res 2005;37:231–45.

[13] Cortell R. Effects of viscous dissipation and work done by deformation on the MHD flow and heat transfer of a viscoelastic fluid over a stretching sheet. Phys Lett A 2006;357:298–305.

[14] Crane LJ. Flow past a stretching plate. Z Angew Math Phys 1970;21:645–7.

[15] Dandapat BS, Santra B, Vajravelu K. The effects of variable fluid properties and thermocapillarity on the flow of a thin film on an unsteady stretching sheet. Int J Heat Mass Transfer 2007;50:991–6.

[16] Gupta PS, Gupta AS. Heat and mass transfer on a stretching sheet with suction or blowing. Can J Chem Eng 1977;55:744–6.

[17] Hayat T, Abbas Z, Sajid M. Series solution for the upper-convected Maxwell fluid over a porous stretching plate. Phys Lett A 2006;358:396–403.

[18] Ingham DB, Pop I, editors. Transport phenomena in porous media. Oxford: Elsevier; 2005, vol. III.

[19] Ishak A. Mixed convection boundary layer flow over a vertical cylinder with prescribed surface heat flux. J Phys A: Math Theor 2009;42:195501.

[20] Kumaran V, Banerjee AK, Kumar AV, Vajravelu K. MHD flow past a stretching permeable sheet. Appl Math Comput 2008;210:26–32.

[21] Liu IC, Andersson HI. Heat transfer in a liquid film on an unsteady stretching sheet. Int J Thermal Sci 2008;47:766–72.

[22] Mukhopadhyay S. Effect of thermal radiation on unsteady mixed convection flow and heat transfer over a porous stretching surface in porous medium. Int J Heat Mass Transfer 2009;52:3261–5.

[23] Mukhopadhyay S. Effects of slip on unsteady mixed convective flow and heat transfer past a stretching surface. Chin Phys Lett 2010;27(12):124401.

[24] Mukhopadhyay S. Heat transfer analysis for unsteady MHD flow past a non-isothermal stretching surface. Nucl Eng Des 2011;241:4835–9.

[25] Mukhopadhyay S. Chemically reactive solute transfer in boundary layer slip flow along a stretching cylinder. Front Chem Sci Eng 2011;5:385–91.

[26] Mukhopadhyay S, Vajravelu K. Effects of transpiration and internal heat generation/absorption on the unsteady flow of a Maxwell fluid at a stretching surface. ASME J Appl Mech 2012;79:http://dx.doi.org/10.1115/1.4006260, 044508-1-6.

[27] Mukhopadhyay S, Vajravelu K, Van Gorder RA. Flow and heat transfer in a moving fluid over a moving non-isothermal surface. Int J Heat Mass Transfer 2012;55(23-24): 6632–7.

[28] Prasad KV, Vajravelu K, Datti PS. Mixed convection heat transfer over a non-linear stretching surface with variable fluid properties. Int J Non-Linear Mech 2010;45:320–30.

[29] Prasad KV, Vajravelu K, Datti PS. The effects of variable fluid properties on the hydromagnetic flow and heat transfer over a non-linearly stretching sheet. Int J Thermal Sci 2010;49:603–10.

[30] Prasad KV, Sujatha A, Vajravelu K, Pop I. MHD flow and heat transfer of a UCM fluid over a stretching surface with variable thermo physical properties. Meccanica 2012;47: 1425–39.

[31] Sajid M, Ahmad I, Hayat T, Ayub M. Unsteady flow and heat transfer of a second grade fluid over a stretching sheet. Comm Nonlinear Sci Numer Simulat 2009;14:96–108.

[32] Schlichting H. Boundary-layer theory. New York: McGraw-Hill; 2008.

[33] Sharidan S, Mahmood T, Pop I. Similarity solutions for the unsteady boundary layer flow and heat transfer due to a stretching sheet. Int J Appl Mech Eng 2006;11:647–54.

[34] Siddiqui AM, Zeb A, Ghori QK, Benharbit AM. Homotopy perturbation method for heat transfer flow of a third grade fluid between parallel plates. Chaos Solitons Fractals 2008;36:182–92.

[35] Soundalgekar VM, Takhar HS, Vighnesam NV. The combined free and forced convection flow past a semi-infinite plate with variable surface temperature. Nucl Eng Des 1988;110: 95–8.

[36] Takhar HS, Gorla RSR, Soundalgekar VM. Radiation effects on MHD free convection flow of a gas past a semi-infinite vertical plate. Int J Numer Meth Heat Fluid Flow 1996;6:77–83.

[37] Tsai R, Huang KH, Huang JS. Flow and heat transfer over an unsteady stretching surface with a non-uniform heat source. Int Comm Heat Mass Tran 2008;35:1340–3.

[38] Vafai K, Tien CL. Boundary and inertia effects on flow and heat transfer in porous media. Int J Heat Mass Transfer 1981;24:195–204.

[39] Vajravelu K, Prasad KV, Chiu-On Ng. Unsteady flow and heat transfer in a thin film of Ostwald–de Waele liquid over a stretching surface. Comm Nonlinear Sci Numer Simulat 2012;17:4163–73.

[40] Van Gorder RA, Vajravelu K. Hydromagnetic stagnation point flow of a second grade fluid over a stretching sheet. Mech Res Comm 2010;37:113–8.

[41] Van Gorder RA, Sweet E, Vajravelu K. Nano boundary layers over stretching surfaces. Comm Nonlinear Sci Numer Simulat 2010;15:1494–500.

[42] Van Gorder RA, Vajravelu K, Akyildiz FT. Existence and uniqueness results for a non-linear differential equation arising in viscous flow over a nonlinearly stretching sheet. Appl Math Lett 2011;24:238–42.

[43] Van Gorder RA, Vajravelu K. Convective heat transfer in a conducting fluid over a permeable stretching surface with suction and internal heat generation/absorption. Appl Math Comput 2011;217:5810–21.

[44] Vleggaar J. Laminar boundary-layer behavior on continuous accelerating surfaces. Chem Eng Sci 1977;32:1517–25.

[45] Wang CY. Flow due to a stretching boundary with partial slip—an exact solution of the Navier-Stokes equations. Chem Eng Sci 2002;57:3745–7.

Part I

Methods and applications

Numerical methods

1

The governing equations of fluid flow problems are generally of a nonlinear and boundary value type. Usually, the exact solutions of the boundary value problems (BVPs) are very difficult to obtain, and so we have to use numerical methods. For some special classes of flow problems, a set of partial differential equations are transformed into a set of ordinary differential equations with the help of similarity variables. The transformed (final) equations then can be solved analytically or numerically. The procedure for finding a numerical solution for a BVP is generally more difficult than that of an initial value problem (IVP). A number of methods can be used to solve linear BVPs. The method of differences is useful in such cases. But this method cannot be used for nonlinear equations. Other methods can be used to obtain linearly independent solutions, which can then be combined in such a way that they satisfy the boundary conditions (Mukhopadhyay [1]). For such problems, the difference method can be adapted. The most popular numerical method is the shooting method. The shooting method can be used for both linear and nonlinear problems. Because the convergence of the method depends on a good initial guess, there is no guarantee that the method will converge. But the method is easy to apply, and when it does converge, it is usually more efficient than other methods (Mukhopadhyay [1], Mukhopadhyay and Layek [2]). Moreover, the shooting method gives more accurate results if guess values (slopes) are chosen correctly.

The basic idea of a shooting method is to replace the BVP by some IVP where the slope at the initial point is obviously unknown. We can guess this unknown quantity, and then, using iteration, the guess value can be improved.

The shooting method consists of the following steps:

1. transformation of the given BVP to an IVP,
2. finding a solution of the IVP, and
3. finding a solution of the given BVP.

Let us consider a nonlinear second-order differential equation $y'' = f\left(x, y, y'\right)$ with the boundary conditions $y(a) = y_0$ and $y(b) = y_1$.

At first, we set $y' = p$ and $p' = f(x, y, p)$ with $y = y_0$ at $x = a$. Actually, the equation is rewritten in terms of a first-order system of two unknown functions. Because these equations are nonlinear, we cannot get the solution by superposition principle. In order to integrate the above system as an IVP, we require a value for p at $x = a$ that is $y'(a)$. Generally, we take a guess for $p(a)$ and use it to obtain a numerical solution. Then comparing the calculated value for y at $x = b$ with the given boundary condition $y = y_1$ at $x = b$ and adjusting the guess value, $p(a)$, we get a better approximation for the solution. The derivative at $x = a$ gives the trajectory of computed solution. That is why, it is called shooting method. Basically, we seek a solution that also satisfies the relation $y(b) = y_1$. A suitable solution can be found by successively refining the interval. With the help of linear interpolation or other root-finding methods, it is also

Fluid Flow, Heat and Mass Transfer at Bodies of Different Shapes. http://dx.doi.org/10.1016/B978-0-12-803733-1.00001-6

possible to improve the obtained solution. To start the integration, the initial slope, that is, $y'(a)$, is required, and the shooting method depends on the choice of the value of $y'(a)$.

To explain the method, we consider the equation

$$f'''(\eta) + \frac{1}{2}f(\eta)f''(\eta) = 0, \tag{1.1}$$

along with the boundary conditions

$$f(0) = 0, \; f'(0) = 0, \tag{1.2}$$

$$f'(\infty) = 1, \tag{1.3}$$

which is a third-order nonlinear BVP.

Now, the key factor is to choose an appropriate and suitable finite value of η as $\eta \to \infty$, say η_∞. Here $\eta = \eta_\infty$ corresponds to the edge of the boundary layer. For computational purposes, η_∞ is to be chosen arbitrarily larger than the boundary layer thickness. The most important factor of the shooting method is to choose an appropriate finite value of η_∞. In order to determine η_∞ for the BVP stated by equations (1.1)–(1.3), we start with some initial guess value α that must be determined so that the resulting solutions yield the prescribed value $f' = 1$ at $\eta = \eta_\infty$ for some particular set of physical parameters (Mukhopadhyay and Layek [2]). We therefore guess at the initial slope, and an iterative procedure is set up for convergence to the correct slope. A normally better approximation to α can now be obtained by the following linear interpolation formula:

$$\alpha_2 = \alpha_0 + (\alpha_1 - \alpha_0)\frac{f'(\eta_\infty) - f'(\alpha_0, \eta_\infty)}{f'(\alpha_1, \eta_\infty) - f'(\alpha_0, \eta_\infty)},$$

where α_0, α_1 are two guesses at the initial slope $f''(0)$ and $f'(\alpha_0, \eta_\infty), f'(\alpha_1, \eta_\infty)$ are the values of f' at $\eta = \eta_\infty$. We now integrate the differential equation using the initial values $f(0) = 0, f'(0) = 0$, and $f''(0) = \alpha_3$ to obtain $f'(\alpha_2, \eta_\infty)$. Using linear interpolation based on α_1, α_2, we can obtain a next approximation α_3. This process is repeated until convergence is obtained. The convergence depends on a good initial guess (Conte and Boor [3]).

The solution procedure is repeated with another large value of η_∞ until two successive values of $f''(0)$ differ only by the specified significant digit. The last value of η_∞ is finally chosen to be the most appropriate value of the limit η_∞. The value of η_∞ may change for another set of physical parameters, if involved in the problem. Once the finite value of η_∞ is determined, then the integration is carried out. We compare the calculated value for f' at $\eta = 15$ (say) with the given boundary condition $f'(15) = 1$, and the estimated value is adjusted using the secant method to get a better approximation for the solution (Mukhopadhyay [4]).

We take the series of values for $f''(0)$ and apply the fourth-order classical Runge-Kutta method with step size $h = 0.01$. The above procedure is repeated until we get the results up to the desired degree of accuracy, 10^{-6}.

A shooting method that can be used to solve this equation or other nonlinear ordinary differential equations was developed by Keller [5]. (For other methods, see Vajravelu and Van Gorder [6], and Vajravelu and Prasad [7]). One of the features of this method is the systematic way in which new values of $f''(0)$ are determined. With the help of Newton's method, the traditional trial-and-error searching technique is replaced (see Isaacson and Keller [8]). Newton's method is a second-order one. Hence, it is more efficient than other methods, for example, the bisection and chord (secant) methods. Newton's method generally provides quadratic convergence of the iterations, and as a result, the computation time decreases. But the main problem of this method is that if the root is too far from the initial value, the iterations do not converge.

According to Keller's shooting method, we first replace equation (1.1) by a system of three first-order ordinary differential equations. If f, f', and f'' are denoted by $f, f_1,$ and f_2, respectively, the system of three first-order equations can be written as

$$f' = f_1, \tag{1.4}$$

$$f_1' = f_2, \tag{1.5}$$

$$f_2' = -\frac{1}{2}ff_2, \tag{1.6}$$

and the boundary conditions are replaced by

$$f(0) = 0, \; f_1(0) = 0, \tag{1.7}$$

$$f_1(\eta_\infty) = 1. \tag{1.8}$$

Let us choose $f_2(0) = s$, s being a guess value. Now we have to find s such that the solution of the IVP satisfies the outer boundary condition (1.8). That is, if we denote the solution of this IVP by $[f(\eta, s), f_1(\eta, s), f_2(\eta, s)]$, then we seek s such that

$$f_1(\eta_\infty, s) - 1 \equiv \phi(s) = 0. \tag{1.9}$$

To solve equation (1.9), we employ Newton's method. This method is widely used for finding the root of an equation by successive approximations. If s^0 is a guess for a root of the equation $\phi(s) = 0$, a better guess, s^1, corresponds to the point of intersection of the s-axis and the tangent of $y = \phi(s)$ at $s = s^0$ and so on. This yields the iteration s^ν defined by

$$s^{\nu+1} = s^\nu - \frac{\phi(s^\nu)}{(d\phi/ds)(s^\nu)} \equiv s^\nu - \frac{f_1(\eta_\infty, s^\nu) - 1}{(\partial f_1/\partial s)(\eta_\infty, s^\nu)}, \quad \nu = 0, 1, 2, \ldots. \tag{1.10}$$

Obviously, $\phi(s^\nu) = 0$, only when $s^{\nu+1} = s^\nu$, and equation (1.8) is satisfied exactly. Otherwise, we iterate until $|s^{\nu+1} - s^\nu| \leq \varepsilon$ for some sufficiently small ε. Then the condition (1.8) is also approximately satisfied.

Now to find the derivative of f_1 with respect to s, we take the derivatives of equations (1.4)–(1.6), and (1.7). This leads to the following linear differential equations, known as the variational equations:

$$F' = U, \tag{1.11}$$

$$U' = V, \tag{1.12}$$

$$V' = -\frac{1}{2}(fV + f_2 F), \tag{1.13}$$

and the initial conditions are

$$F(0) = 0, \; U(0) = 0, \; V(0) = 1, \tag{1.14}$$

where $F(\eta, s) \equiv \dfrac{\partial f}{\partial s}$, $U(\eta, s) \equiv \dfrac{\partial f_1}{\partial s}$, $V(\eta, s) \equiv \dfrac{\partial f_2}{\partial s}$.

Note that the boundary condition at η_∞ has disappeared in (1.14), that is, the BVP given by the equations (1.4)–(1.6) and by (1.7) is transformed to the IVP given by the equations (1.11)–(1.14).

Now, we use a fourth-order Runge-Kutta method to solve the IVP. It is a "self-starting" method. Other, non-self-starting methods require details of the solution for several previous steps before a new step can be executed by integrating polynomial fits to the previous values of the derivatives. However, the Runge-Kutta method can "pull itself up by its bootstraps." As might be excepted, non-self-starting methods are faster to run, and where many calculations are to be done, it is common to start with a Runge-Kutta method and then switch over to a non-self-starting method such as the predictor-corrector method described by Isaacson and Keller [6].

References

[1] Mukhopadhyay S. Mixed convection boundary layer flow along a stretching cylinder in porous medium. J Petrol Sci Eng 2012;96–97:73–8.
[2] Mukhopadhyay S, Layek GC. Radiation effect on forced convective flow and heat transfer over a porous plate in a porous medium. Meccanica 2009;44:587–97.
[3] Conte SD, Boor C. Elementary numerical analysis. New York: McGraw-Hill; 1981, 412.
[4] Mukhopadhyay S. Slip effects on MHD boundary layer flow over an exponentially stretching sheet with suction/blowing and thermal radiation. Ain Shams Eng J 2013;4:485–91.
[5] Keller HB. Numerical methods in boundary-layer theory. Annu Rev Fluid Mech 1968;10:417–33.
[6] Vajravelu K, Van Gorder RA. Nonlinear flow phenomena and homotopy analysis: fluid flow and heat transfer. Beijing: Higher Education Press; 2012, and Springer-Verlag, Berlin.
[7] Vajravelu K, Prasad KV. Keller-Box method and its application (de Gruyter studies in mathematical physics). Berlin/Boston: HEP - de Gruyter GmbH; 2014.
[8] Isaacson E, Keller HB. Analysis of numerical methods. John Wiley; 1966.

Flow past a stretching sheet

2

Almost a century ago, Prandtl realized that the boundary layer plays a significant role in determining accurately the flow of certain fluids. He showed that for slightly viscous fluid, viscosity is negligible in the bulk of the flow but it assumes a vital role near the boundaries. The asymptotic procedure to derive the boundary layer equations was first introduced by Prandtl, which considerably changed the mathematical properties of the original Navier-Stokes equations. These equations are of the elliptic type whereas the boundary layer equations are of the parabolic type. The parabolic nature of the boundary layer equations makes the mathematical treatment substantially easier in order to obtain numerical solutions for practical problems.

The quest for exact solutions of the boundary layer equations has a long history. Exact solutions play an important role in hydrodynamics because they can be used as a basis for description of complex motions of liquids and gases. General methods for finding exact solutions of linear partial differential equations indeed exist. The nonlinear character of the partial differential equations governing the motion of a fluid produces difficulties in solving the equations. A self-similar equation and its solutions are useful for the analysis of some flow problems. In the field of fluid mechanics, most of the researchers try to obtain the similarity solutions in such cases.

The study of laminar flow and heat transfer over a stretching sheet is of considerable interest because of its ever-increasing industrial applications and important bearings in several technological processes (Mukhopadhyay [1]). Boundary layer flow over a stretching surface is often encountered in many engineering disciplines. It is generally assumed that the sheet is inextensible. But many cases arise in polymer industry in which it is necessary to deal with a stretching sheet (see [1]). The production of sheeting material arises in a number of industrial manufacturing processes and includes both metal and polymer sheets. In the manufacture of the latter, the material is in a molten phase when thrust through an extrusion die and then cools and solidifies some distance away from the die before arriving at the collecting stage (Mukhopadhyay and Gorla [2]). The tangential velocity imported by the sheet induces motion in the surrounding fluid, which alters the convection cooling of the sheet. Similar situations prevail during the manufacture of plastic and rubber sheets, where it is often necessary to blow a gaseous medium through the not-yet solidified material, and where the stretching force may be varying with time (see [2]). Another example that belongs to this class of problems is the cooling of a large metallic plate in a bath, which may be an electrolyte. Glass blowing, continuous casting, and spinning of fibers also involve the flow due to a stretching surface. Because of the much higher viscosity of the fluid near the sheet, one can assume that the fluid is affected by the sheet but not vice versa (Mukhopadhyay and Gorla [2]). Thus, the fluid dynamic problem can be idealized to the case of a fluid disturbed by a tangentially moving boundary. Experiments show that the velocity of the boundary is approximately proportional to the distance from the orifice. The quality of the resulting sheeting material, as well as

Fluid Flow, Heat and Mass Transfer at Bodies of Different Shapes. http://dx.doi.org/10.1016/B978-0-12-803733-1.00002-8

the cost of production, is affected by the speed of collection and the heat transfer rate, and knowledge of the flow properties of the ambient fluid is clearly desirable (see [2]).

In this chapter, we shall discuss the heat and mass transfer characteristics for flow over a stretching surface. Most of the available literature deals with the study of boundary layer flow over a linearly stretching surface. However, realistically, stretching of plastic sheet may not necessarily be linear (Gupta and Gupta [3]). A number of real processes are thus undertaken using different stretching velocities such as linear, power-law, and exponential. In Section 2.1, we shall review the flow and heat transfer characteristics for flow past a linearly stretching sheet, whereas flow past a nonlinearly stretching sheet is outlined in Section 2.2. We are then able to outline the mathematical formulation and the numerical solution to the nonlinear differential equations and relevant boundary conditions arising in the problem of flow past an exponentially stretching surface in Section 2.3. In all these sections, the flow and temperature fields are considered to be at steady state. However, in some cases, the flow field and heat and mass transfer can be unsteady because of a sudden stretching of the flat sheet or by a step change of the temperature of the sheet. Accordingly, in Section 2.4, we discuss the situation when the stretching force and surface temperature are varying with time. Though there are often multiple ways to solve a given problem, the similarity method is adopted here to transform the governing time-dependent boundary layer equations into a set of ordinary differential equations and then the numerical solutions are obtained. In all these sections, we consider that the stretching surface is flat. In reality, the stretching surface may not be flat. The flow due to a curved stretching surface may have applications in a stretch-forming machine with curving jaws. Hence, in Section 2.5, we present a theoretical study to obtain a numerical solution for the viscous flow over a curved stretching surface. So far, we have discussed the flow and heat and mass transfer characteristics of Newtonian fluid past a stretching sheet. Recently, dynamics of non-Newtonian fluid has become quite a popular topic of research interest. Such interest in fact stems from the applications of these fluids in biology, physiology, technology, and industry. Keeping this in mind, in Section 2.6, stagnation point flow of non-Newtonian fluid over a stretching surface is analyzed.

2.1   Flow past a linearly stretching sheet

Boundary layer flow over a stretching surface has received great attention owing to its applications in many engineering disciplines. This type of flow occurs in various areas such as a cooling bath, the boundary layer along a material-handling conveyer, the boundary layer along liquid film and condensation processes, the cooling or drying of papers and textiles, glass fiber production, etc. (Prasad et al. [4]). In particular, in plastic and metal industries, the flow induced by a stretching surface is important in the extrusion processes (Zheng et al. [5]). A stretched sheet interacts both thermally and mechanically with the ambient fluid during the manufacturing process (Pal [6]). Actually, stretching imparts unidirectional orientation to the extrudate. Consequently, the quality of the final product depends considerably on the flow and heat transfer mechanism (Vajravelu et al. [7]). Therefore, the analysis of momentum and thermal transport in the case of a continuously stretching surface is important for gaining some fundamental understanding of such processes. The tangential velocity imparted by the surface induces motion in the surrounding fluid, and this alters the convection cooling of the surface (Mukhopadhyay et al. [8]). Knowledge of the flow properties of the fluid is desirable because the quality of the resulting sheeting material, as well as the cost of production, is affected by the speed of collection. Therefore, the flow behavior over a stretching surface that determines the rate of cooling is an important aspect nowadays (Mukhopadhyay et al. [8], Mahapatra et al. [9]). Crane [10] investigated the steady flow of an incompressible viscous fluid over an elastic sheet, the flow being caused solely by the stretching of the sheet in its own plane with a velocity varying linearly with the distance from a fixed point. He obtained the closed-form solution of the problem. Heat transfer in the flow over a stretching surface maintained at constant as well as variable temperature was investigated by Carragher and Crane [11] and Gupta and Gupta [3]. Chen and Char [12] studied the heat transfer characteristics on a stretching sheet with variable surface temperature in the case of suction or blowing. In recent years, Crane's [10] problem has been extended by many researchers by including various aspects of flow and heat transfer characteristics when the surface is stretched with a linear velocity. For details, the readers are referred to Dutta et al. [13], Vajravelu [14], Mukhopadhyay et al. [15], Ishak et al. [16], and many others.

In most of the studies in literature, the thermophysical properties of the ambient fluids were assumed to be constant. However it is well known that these properties may change with temperature, especially the thermal conductivity. Most fluids have temperature-dependent properties and, under circumstances where large or moderate temperature gradients exist across the fluid medium, fluid properties often vary significantly. It is known from physics that with the rise of temperature, the coefficient of viscosity decreases in case of liquids, whereas it increases in case of gases. In Mukhopadhyay et al. [17], the authors considered the viscosity model given by Batchelor [18] describing the viscous flow due to a stretching surface in the presence of an applied magnetic field.

A similarity transformation is generally applied in such cases to convert the governing partial differential equations into nonlinear ordinary differential equations. Here in this section we shall present the results obtained by Mukhopadhyay et al. [17] with slight modifications, where the scaling group of transformations, a special form of Lie group transformations, was employed to find the similarity solutions of the relevant boundary value problem. This is the only mathematical method to find the full set of symmetries of a given differential equation and no ad hoc assumptions or prior knowledge of that equation are needed (Mukhopadhyay and Layek [19]).

We also obtain numerical solutions by converting the boundary value problem into an initial value problem first and then obtaining solutions via the Runge-Kutta method.

2.1.1 Mathematical analysis of the problem

Let (u, v) be the velocity components in the (x, y) directions, respectively. Then the continuity, momentum, and energy equations for steady two-dimensional flow of a viscous incompressible electrically conducting fluid past a heated linearly stretching sheet in presence of a uniform magnetic field of strength B_0 (imposed along the y-axis) can be written as

$$\frac{\partial u}{\partial x} + \frac{\partial v}{\partial y} = 0,$$
(2.1)

$$u\frac{\partial u}{\partial x} + v\frac{\partial u}{\partial y} = \frac{1}{\rho}\frac{\partial \mu}{\partial T}\frac{\partial T}{\partial y}\frac{\partial u}{\partial y} + \frac{\mu}{\rho}\frac{\partial^2 u}{\partial y^2} - \frac{\sigma B_0^2}{\rho}u,$$
(2.2)

$$u\frac{\partial T}{\partial x} + v\frac{\partial T}{\partial y} = \frac{\partial}{\partial y}\left(\kappa\frac{\partial T}{\partial y}\right).$$
(2.3)

Here T is the temperature, κ is the thermal diffusivity, ρ is the fluid density (assumed constant), μ is the coefficient of fluid viscosity, and σ is the conductivity of the fluid. Because the variation of thermal conductivity (hence thermal diffusivity) and viscosity μ with temperature is quite significant, the thermal diffusivity κ and the fluid viscosity are assumed to vary with temperature. In the range of temperature considered (i.e. $0-23°C$), the variation of both density ρ and specific heat c_p with temperature is negligible. Hence, they are taken as constants.

As in Mukhopadhyay et al. [17], we take the boundary conditions as

$$u = cx, v = v_w, T = T_w \text{ at } y = 0,$$
(2.4a)

$$u \to \infty, T \to T_\infty \text{ as } y \to \infty,$$
(2.4b)

where $c > 0$ is the stretching rate of the sheet, v_w is the velocity at the wall where $v_w < 0$ corresponds to the case of suction and $v_w > 0$ for blowing. T_w is the uniform wall temperature and T_∞ is the free stream temperature.

We introduce the relations

$$u = \frac{\partial \psi}{\partial y}, v = -\frac{\partial \psi}{\partial x}, \theta = \frac{T - T_\infty}{T_w - T_\infty},$$
(2.5)

where ψ is the stream function and θ is the dimensionless temperature.

Temperature-dependent fluid viscosity is given by (Batchelor [18])

$$\mu = \mu^*[a + b(T_w - T)]$$
(2.6)

where μ^* is the constant value of the coefficient of viscosity far away from the sheet, a, b are constants with $b > 0$.

The variation of thermal diffusivity with the dimensionless temperature is written as $\kappa = \kappa_0(1 + \lambda\theta)$, where λ is a parameter that depends on the nature of the fluid and κ_0 is the value of thermal diffusivity at the temperature T_w. This relation agrees well with that of Saikrishnan and Roy [20] and also with Bird et al. [21] (neglecting second- and higher-order terms).

With the help of the relations (2.5) and (2.6), the equations (2.1)–(2.3) can be written as

$$\frac{\partial \psi}{\partial y}\frac{\partial^2 \psi}{\partial x \partial y} - \frac{\partial \psi}{\partial x}\frac{\partial^2 \psi}{\partial y^2} = -A\upsilon^*\frac{\partial \theta}{\partial y}\frac{\partial^2 \psi}{\partial y^2} + \upsilon^*[a + A(1-\theta)]\frac{\partial^3 \psi}{\partial y^3} - cM^2\frac{\partial \psi}{\partial y}, \qquad (2.7)$$

$$\frac{\partial \psi}{\partial y}\frac{\partial \theta}{\partial x} - \frac{\partial \psi}{\partial x}\frac{\partial \theta}{\partial y} = \kappa_0(1+\lambda\theta)\frac{\partial^2 \theta}{\partial y^2} + \kappa_0\lambda\left(\frac{\partial \theta}{\partial y}\right)^2, \qquad (2.8)$$

where $A = b(T_w - T_\infty)$, $\upsilon^* = \dfrac{\mu^*}{\rho}$ and $\dfrac{\sigma B_0^2}{\rho} = cM^2$, M being the Hartman number.

Boundary conditions become

$$\frac{\partial \psi}{\partial y} = cx,\ \frac{\partial \psi}{\partial x} = -\upsilon_w, \theta = 1 \quad \text{at}\ \ y = 0, \qquad (2.9a)$$

$$\frac{\partial \psi}{\partial y} \to 0, \theta \to 0 \quad \text{as}\ \ y \to \infty. \qquad (2.9b)$$

Introducing the simplified form of Lie group transformations, namely, the scaling group of transformations as in Mukhopadhyay et al. [17],

$$\Gamma: x^* = xe^{\varepsilon\alpha_1}, y^* = ye^{\varepsilon\alpha_2}, \psi^* = \psi e^{\varepsilon\alpha_3}, u^* = ue^{\varepsilon\alpha_4}, v^* = ve^{\varepsilon\alpha_5}, \theta^* = \theta e^{\varepsilon\alpha_6}. \qquad (2.10)$$

The transformation Γ transforms the point $(x, y, \psi, u, v, \theta)$ to $(x^*, y^*, \psi^*, u^*, v^*, \theta^*)$. Substituting (2.10) into equations (2.7) and (2.8), we get

$$e^{\varepsilon(\alpha_1 + 2\alpha_2 - 2\alpha_3)}\left(\frac{\partial \psi^*}{\partial y^*}\frac{\partial^2 \psi^*}{\partial x^*\partial y^*} - \frac{\partial \psi^*}{\partial x^*}\frac{\partial^2 \psi^*}{\partial y^{*2}}\right) = -A\upsilon^* e^{\varepsilon(3\alpha_2 - \alpha_3 - \alpha_6)}\frac{\partial \theta^*}{\partial y^*}\frac{\partial^2 \psi^*}{\partial y^{*2}}$$

$$+ \upsilon^*(a+A)e^{\varepsilon(3\alpha_2 - \alpha_3)}\frac{\partial^3 \psi^*}{\partial y^{*3}} - \upsilon^* A e^{\varepsilon(3\alpha_2 - \alpha_3 - \alpha_6)}\theta^*\frac{\partial^3 \psi^*}{\partial y^{*3}} - cM^2 e^{\varepsilon(\alpha_2 - \alpha_3)}\frac{\partial \psi^*}{\partial y^*},$$

$$(2.11)$$

$$e^{\varepsilon(\alpha_1 + \alpha_2 - \alpha_3 - \alpha_6)}\left(\frac{\partial \psi^*}{\partial y^*}\frac{\partial \theta^*}{\partial x^*} - \frac{\partial \psi^*}{\partial x^*}\frac{\partial \theta^*}{\partial y^*}\right) = \kappa_0(1 + \lambda\theta^* e^{-\varepsilon\alpha_6})e^{\varepsilon(2\alpha_2 - \alpha_6)}\frac{\partial^2 \theta^*}{\partial y^{*2}}$$

$$+ \kappa_0\lambda e^{2\varepsilon(\alpha_2 - \alpha_6)}\left(\frac{\partial \theta^*}{\partial y^*}\right)^2 \qquad (2.12)$$

The system remains invariant under the transformation Γ if and only if the following relations are satisfied:

$$\alpha_1 + 2\alpha_2 - 2\alpha_3 = 3\alpha_2 - \alpha_3 - \alpha_6 = 3\alpha_2 - \alpha_3 = 3\alpha_2 - \alpha_3 - \alpha_6 = \alpha_2 - \alpha_3, \qquad (2.13a)$$

$$\alpha_1 + \alpha_2 - \alpha_3 - \alpha_6 = 2\alpha_2 - \alpha_6 = 2(\alpha_2 - \alpha_6) \qquad (2.13b)$$

These relations give $\alpha_2 = 0 = \alpha_6$, $\alpha_1 = \alpha_3$.

The boundary conditions yield $\alpha_1 = \alpha_4, \alpha_5 = 0$.
Thus, Γ reduces to

$$\Gamma: x^* = xe^{\varepsilon\alpha_1}, y^* = y, \psi^* = \psi e^{\varepsilon\alpha_1}, u^* = ue^{\varepsilon\alpha_1}, v^* = v, \theta^* = \theta. \tag{2.14}$$

Expanding by Taylor's series, we have

$$x^* - x = x\varepsilon\alpha_1, y^* - y = 0, \psi^* - \psi = \psi\varepsilon\alpha_1,$$
$$u^* - u = u\varepsilon\alpha_1, v^* - v = 0, \theta^* - \theta = 0.$$

In terms of differentials, we have

$$\frac{dx}{\alpha_1 x} = \frac{dy}{0} = \frac{d\psi}{\alpha_1 \psi} = \frac{du}{\alpha_1 u} = \frac{dv}{0} = \frac{d\theta}{0}. \tag{2.15}$$

From the subsidiary equations (2.15), one can easily get

$$\eta = y, \psi = xF(\eta), \theta = \theta(\eta), \tag{2.16}$$

where F is an arbitrary function of η.
Equations (2.11) and (2.12) now become

$$F'^2 - FF'' = -Av^*\theta' F'' + v^*[a + A(1 - \theta)]F''' - cM^2 F', \tag{2.17}$$

$$F\theta' + \kappa_0\lambda\theta'^2 + \kappa_0(1 + \lambda\theta)\theta'' = 0. \tag{2.18}$$

The boundary conditions now become

$$F'(\eta) = c, F(\eta) = S, \theta(\eta) = 1 \text{ at } \eta = 0, \tag{2.19a}$$

$$F'(\eta) \to 0, \theta(\eta) \to 0 \text{ as } \eta \to \infty. \tag{2.19b}$$

Substituting $\eta = v^{*\alpha}c^\beta\eta^*, F = v^{*\alpha'}c^{\beta'}F^*, \theta = v^{*\alpha''}c^{\beta''}\theta^*$ in equations (2.17) and (2.18), we get

$$\alpha = \alpha' = \frac{1}{2}, \alpha'' = 0, \beta' = -\beta = \frac{1}{2}, \beta'' = 0.$$

Taking $\eta^* = \eta, F^* = f, \theta^* = \theta$, the above equations (2.17) and (2.18) and the boundary conditions (2.19a) and (2.19b) become

$$f'^2 - ff'' = -A\theta'f'' + [a + A(1 - \theta)]f''' - M^2 f', \tag{2.20}$$

$$Prf\theta' + \lambda\theta'^2 + (1 + \lambda\theta)\theta'' = 0, \tag{2.21}$$

$$f'(\eta) = 1, f(\eta) = S, \theta(\eta) = 1 \text{ at } \eta = 0, \tag{2.22a}$$

$$f'(\eta) \to 0, \theta(\eta) \to 0 \text{ as } \eta \to \infty. \tag{2.22b}$$

Here, $Pr = \dfrac{\nu^*}{\kappa_0}$ is the Prandtl number, $S = -\dfrac{v_w}{\sqrt{\nu^* c}}$, $S > 0$ ($v_w < 0$) corresponds to suction, and whereas $S < 0$ (i.e. $v_w > 0$) corresponds to blowing.

2.1.2 Numerical results and discussion

We obtain numerical solutions to the above boundary value problem where equations (2.20) and (2.21) are solved as an initial value problem via the Runge-Kutta method. The numerical results are obtained for a region $0 < \eta < \eta_\infty$, where $\eta_\infty = 7$. In all the numerical computations, we have taken $a = 1$.

In order to analyze the results, numerical computation has been carried out using the shooting method for various values of the temperature-dependent fluid viscosity parameter A and variable thermal diffusivity parameter λ.

First, we concentrate on the effects of temperature-dependent fluid viscosity on velocity distribution and heat transfer. From Fig. 2.1(a), it is noted that the velocity field is found to decay with the increasing value of η for all the values of A considered. Fluid velocity is found to increase with the increasing values of A (Fig. 2.1(a)). Such effects are not significant in the near wall region. As we go away, it can be seen easily that these effects become more and more significant with the increase of A. But this feature is found to be reduced again far away from the wall and smeared out finally at about $\eta = 10$. Temperature is found to decrease with the increase of η until it vanishes at $\eta = 10$ (Fig. 2.1(b)). But the temperature is found to decrease for any nonzero fixed value of η with the increase of A. It may be easily seen that at any location, the value of θ becomes lower and lower with the increase of the parameter A. An increase in the temperature-dependent fluid viscosity parameter A decreases the thermal boundary layer thickness, which results in a decrease of temperature profile $\theta(\eta)$.

Figure 2.2(a) and (b) exhibit the nature of horizontal velocity and temperature fields, respectively, with variable thermal diffusivity parameter λ. Horizontal velocity is found to decrease with increasing values of λ (Fig. 2.2(a)), whereas the temperature at a particular point of the sheet increases with increasing values of λ (Fig. 2.2(b)). This is due to the thickening of the thermal boundary layer as a result of increasing thermal diffusivity. So the horizontal velocity decreases.

2.2 Flow past a nonlinearly stretching sheet

Flow due to a stretching sheet is very important in the extrusion of sheet materials. Most of the available literature deals with the study of boundary layer flow over a stretching surface where the velocity of the stretching surface is assumed linearly proportional to the distance from the fixed origin (Mukhopadhyay [22]). However, a stretching surface may not be linear. Various aspects of such problems have been investigated by many authors [23–27]. Akyildiz et al. [28] consider the velocity $u = cx^n$ at $y = 0$, which was employed for positive odd integer values of n. It is clear that such

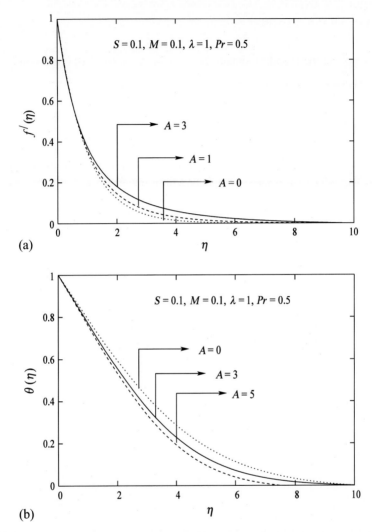

Figure 2.1 Effects of variable viscosity parameter A on (a) velocity and (b) temperature profiles.

a profile would fail for even integer values of n, as the flow at $y = 0$ would be in the wrong direction in the case of $-\infty < x < 0$ (see Van Gorder and Vajravelu [29]). With the help of the modification provided in Van Gorder and Vajravelu [29], one can account for any values of $n \geq 1$, even nonintegers. This allows us to consider a more general nonlinear power-law stretching of the sheet.

The authors of the studies mentioned above continued their discussions by assuming the no-slip boundary conditions. The nonadherence of the fluid to a solid boundary, also known as velocity-slip, is a phenomenon that has been observed under certain

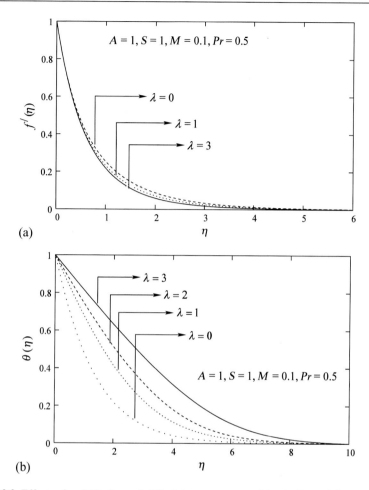

Figure 2.2 Effects of variable thermal diffusivity parameter on (a) velocity and (b) temperature profiles.

circumstances (Yoshimura and Prudhomme [30]). To describe the phenomenon of slip, Navier [31] introduced a boundary condition which states that the component of the fluid velocity tangential to the boundary walls is proportional to the tangential stress (Mukhopadhyay et al. [32]). Recently, many researchers (Wang [33], Andersson [34], Ariel et al. [35], Ariel [36], Abbas et al. [37], Mukhopadhyay and Andersson [38], Mukhopadhyay [39], Mukhopadhyay and Gorla [1]) investigated the flow problems using slip flow conditions at the boundary. Mukhopadhyay [22] extended the problem discussed by Van Gorder and Vajravelu [29] by considering the slip at the boundary. The existence and uniqueness of the results for the problem were presented in Van Gorder and Vajravelu [29] along with some numerical results. In this section, we shall present the results obtained by Mukhopadhyay [22].

2.2.1 Formulation of the problem

Let (u, v) be the velocity components in the (x, y) directions, respectively. Then the continuity and momentum equations governing the flow of an incompressible viscous fluid past a nonlinearly stretching sheet can be written as

$$\frac{\partial u}{\partial x} + \frac{\partial v}{\partial y} = 0, \tag{2.23}$$

$$u\frac{\partial u}{\partial x} + v\frac{\partial u}{\partial y} = \nu\frac{\partial^2 u}{\partial y^2}, \tag{2.24}$$

where $\nu = \frac{\mu}{\rho}$ is the kinematic viscosity, ρ is the fluid density, μ is the coefficient of fluid viscosity.

Polymer melts often exhibit macroscopic wall slip and that in general is governed by a nonlinear and monotone relation between the slip velocity and traction. According to Gad-el-Hak [40], velocity slip is assumed to be proportional to local shear stress.

The appropriate boundary conditions for the problem are given by

$$u = c\operatorname{sgn}(x)|x|^M + N\nu\frac{\partial u}{\partial y}(-\infty < x < \infty), v = -V(x) \quad \text{at} \quad y = 0, \tag{2.25a}$$

$$u \to 0 \quad \text{as} \quad y \to \infty. \tag{2.25b}$$

Here, $c\ (>0)$ is a constant; $M\ (>0)$ is a nonlinear stretching parameter; $N = N_1|x|^{-\frac{M-1}{2}}$ is the velocity slip factor, which changes with x; and N_1 is the initial value of velocity slip factor. The no-slip case is recovered for $N = 0$. $V(x) > 0$ is the velocity of suction and $V(x) < 0$ is the velocity of blowing. $V(x) = V_0\operatorname{sgn}(x)|x|^{\frac{M-1}{2}}$, a special type of velocity at the wall is considered.

We introduce the following similarity variable and similarity transformations:

$$\eta = y\sqrt{\frac{c(M+1)}{2\nu}}|x|^{\frac{M-1}{2}}, \quad u = c\operatorname{sgn}(x)|x|^M f'(\eta),$$

$$v = -\operatorname{sgn}(x)\sqrt{\frac{c(M+1)\nu}{2}}|x|^{\frac{M-1}{2}}\left\{f(\eta) + \left(\frac{M-1}{M+1}\right)\eta f'(\eta)\right\} \tag{2.26}$$

and upon substitution of (2.26) in equations (2.24), (2.25a), and (2.25b), the governing equations and the boundary conditions reduce to

$$f''' + ff'' - \frac{2M}{M+1}f'^2 = 0, \tag{2.27}$$

$$f' = 1 + Bf'', f = S \quad \text{at} \quad \eta = 0 \tag{2.28a}$$

and

$$f' \to 0 \quad \text{as} \quad \eta \to \infty \tag{2.28b}$$

where the prime denotes differentiation with respect to η, $B = N_1 \sqrt{\frac{c\nu(M+1)}{2}}$ is the slip parameter, and $S = \dfrac{V_0}{\sqrt{\frac{c\nu(M+1)}{2}}} > 0$ (or < 0) is the suction (or blowing) parameter.

2.2.2 Numerical solutions and discussion of the results

Numerical computations have been carried out using the fourth-order classical Runge-Kutta method with shooting technique.

In Fig. 2.3(a), velocity profiles are shown for different values of the nonlinear stretching parameter M. The velocity curves show that the rate of transport is considerably reduced with increasing values of M. Further, the decrease in $f'(\eta)$ is almost negligible for large M as the coefficient $2M/(M+1)$ in the differential equation (2.27) approaches 2 as M approaches infinity.

With increasing B, velocity is found to decrease initially (Fig. 2.3(b)). This feature prevails up to certain heights and then the process is slowed down, and at a far distance from the wall, fluid velocity increases slightly.

When slip occurs, the flow velocity near the sheet is no longer equal to the stretching velocity of the sheet. With increase in B, slip velocity increases and consequently fluid velocity decreases because under the slip condition the pulling of the stretching sheet can be only partly transmitted to the fluid.

Figure 2.4(a) exhibits the nature of the skin-friction coefficient with nonlinear stretching parameter M for slip ($B = 0.6$) and no slip ($B = 0$) cases. It is found that skin-friction coefficient increases with M whereas it decreases in the presence of slip at the boundary. Figure 2.4(b) displays the behavior of the skin-friction coefficient against the suction parameter S for a nonlinearly stretching sheet for two values of the slip parameter B. The skin-friction coefficient decreases with slip and increases with suction.

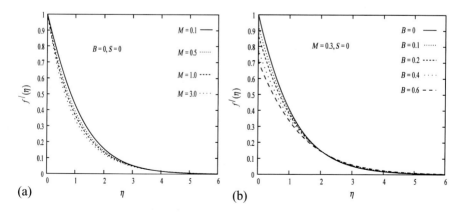

Figure 2.3 Effects of (a) nonlinear stretching parameter and (b) slip parameter on velocity profiles [22].

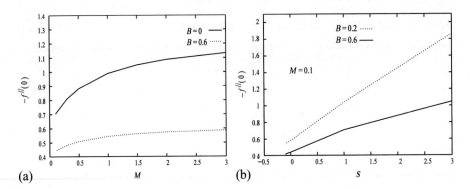

Figure 2.4 Skin friction coefficient against (a) nonlinearly stretching and (b) suction parameter [22].

2.3   Flow past an exponentially stretching sheet

Several researchers investigated laminar flow and heat transfer past a stretching sheet because of its ever-increasing industrial applications and important bearings on several technological processes. Most of the earlier investigations deal with the study of boundary layer flow past a stretching surface in which the velocity of the stretching surface is assumed linearly proportional to the distance from the fixed origin. Vajravelu [24] reported that a stretching surface may not necessarily be linear.

Flow and heat transfer characteristics past an exponentially stretching sheet have a wide variety of applications in technology. For example, in case of annealing and thinning of copper wires, the final product depends on the rate of heat transfer at the stretching continuous surface with exponential variations of stretching velocity and temperature distribution. During such processes, both the kinematics of stretching and the simultaneous heating or cooling have a decisive influence on the quality of the final products.

The heat and mass transfer on boundary layer flow due to an exponential continuous stretching sheet was considered by Magyari and Keller [41]. Elbashbeshy [42] added a new dimension by considering an exponentially continuous porous stretching surface. The viscous-elastic boundary layer flow and heat transfer due to an exponential stretching sheet was investigated by Khan [43] and Sanjayanand and Khan [44]. Later on, the influence of thermal radiation on the boundary layer flow due to an exponentially stretching sheet was discussed by Sajid and Hayat [45]. They solved the problem analytically using the homotopy analysis method (HAM). Recently, Bidin and Nazar [46] analyzed the effect of thermal radiation on the steady boundary layer flow and heat transfer past an exponentially stretching sheet with the help of numerical solutions. El-Aziz [47] described the flow and heat transfer past an exponentially stretching sheet. The mixed convection flow over an exponentially stretching surface in presence of a magnetic field was investigated by Pal [48]. Recently, Ishak [49] analyzed the magnetic effects on flow and heat transfer past an exponentially stretching surface.

All of the previously mentioned studies assumed no-slip boundary conditions. In no-slip flow, as a requirement of continuum physics, the flow velocity is zero at a solid–fluid interface and the fluid temperature adjacent to the solid walls is equal to that of the solid walls. The nonadherence of the fluid to a solid boundary is known as velocity slip. It is a phenomenon that has been observed under certain circumstances (Bhattacharyya et al. [50]). The fluids exhibiting boundary slip find applications in technology such as in the polishing of artificial heart valves and internal cavities. Recently, using slip flow conditions at the boundary, many researchers investigated the different flow problems over a stretching sheet.

In this section, we shall present the results obtained by Mukhopadhyay [1]. It is hoped that the results obtained will provide useful information for applications as well as being a complement to the previous studies.

2.3.1 Mathematical formulation

Consider the flow of an incompressible viscous fluid past a flat sheet coinciding with the plane $y = 0$. The flow is confined to $y > 0$. Two equal and opposite forces are applied along the x-axis so that the wall is stretched keeping the origin fixed. A variable magnetic field $B(x) = B_0 e^{\frac{x}{2L}}$ is applied normal to the sheet, B_0 being a constant. In physics and engineering, the radiative effects have important applications. In space technology and high-temperature processes, the radiation heat transfer effects on different flows are very important. The continuity, momentum, and energy equations in the presence of thermal radiation can be written as

$$\frac{\partial u}{\partial x} + \frac{\partial v}{\partial y} = 0, \tag{2.29}$$

$$u\frac{\partial u}{\partial x} + v\frac{\partial u}{\partial y} = \nu \frac{\partial^2 u}{\partial y^2} - \frac{\sigma B^2}{\rho} u, \tag{2.30}$$

$$u\frac{\partial T}{\partial x} + v\frac{\partial T}{\partial y} = \frac{\kappa}{\rho c_p}\frac{\partial^2 T}{\partial y^2} - \frac{1}{\rho c_p}\frac{\partial q_r}{\partial y} \tag{2.31}$$

where u and v are the components of velocity respectively in the x and y directions, $\nu = \frac{\mu}{\rho}$ is the kinematic viscosity, ρ is the fluid density, μ is the coefficient of fluid viscosity, σ is the electrical conductivity, q_r is the radiative heat flux, c_p is the specific heat at constant pressure, and κ is the thermal conductivity of the fluid.

Using Rosseland approximation for radiation (Brewster [51]), we can write

$$q_r = -\frac{4\sigma^*}{3k^*}\frac{\partial T^4}{\partial y} \tag{2.31a}$$

where σ^* is the Stefan-Boltzman constant and k^* is the absorption coefficient.

Assuming that the temperature difference within the flow is such that T^4 may be expanded in a Taylor series and expanding T^4 about T_∞ and neglecting higher orders, we get $T^4 \equiv 4T_\infty^3 T - 3T_\infty^4$. Therefore, the equation (2.31) becomes

$$u\frac{\partial T}{\partial x} + v\frac{\partial T}{\partial y} = \frac{\kappa}{\rho c_p}\frac{\partial^2 T}{\partial y^2} + \frac{16\sigma^* T_\infty^3}{3\rho c_p k^*}\frac{\partial^2 T}{\partial y^2}. \tag{2.32}$$

The flow field can be changed significantly by the application of suction or injection (blowing) of a fluid through the bounding surface. In the design of thrust bearing and radial diffusers and thermal oil recovery, the process of suction or blowing is very important.

The appropriate boundary conditions for the problem are given by

$$u = U + N\nu \frac{\partial u}{\partial y}, v = -V(x), T = T_w + D\frac{\partial T}{\partial y} \quad \text{at} \quad y = 0, \tag{2.33a}$$

$$u \to 0, T \to 0 \quad \text{as} \quad y \to \infty. \tag{2.33b}$$

Here $U = U_0 e^{\frac{x}{L}}$ is the stretching velocity; $T_w = T_0 e^{\frac{x}{2L}}$ is the temperature at the sheet; U_0 and T_0 are the reference velocity and temperature, respectively; $N = N_1 e^{-\frac{x}{2L}}$ is the velocity slip factor, which changes with x; N_1 is the initial value of velocity slip factor; $D = D_1 e^{-\frac{x}{2L}}$ is the thermal slip factor which also changes with x; and D_1 is the initial value of thermal slip factor. The no-slip case is recovered for $N = 0 = D$. $V(x) > 0$ is the velocity of suction and $V(x) < 0$ is the velocity of blowing. $V(x) = V_0 e^{\frac{x}{2L}}$, a special type of velocity at the wall, is considered. V_0 is the initial strength of suction.

Introducing the similarity variable and similarity transformations

$$\eta = \sqrt{\frac{U_0}{2\nu L}} e^{\frac{x}{2L}} y, u = U_0 e^{\frac{x}{L}} f'(\eta),$$

$$v = -\sqrt{\frac{\nu U_0}{2L}} e^{\frac{x}{2L}} \{f(\eta) + \eta f'(\eta)\}, T = T_0 e^{\frac{x}{2L}} \theta(\eta), \tag{2.34}$$

upon substitution of (2.34) in equations (2.30) and (2.32), the governing equations reduce to

$$f''' + ff'' - 2f'^2 - M^2 f' = 0, \tag{2.35}$$

$$\left(1 + \frac{4}{3}R\right)\theta'' + Pr\left(f\theta' - f'\theta\right) = 0 \tag{2.36}$$

and the boundary conditions take the following form

$$f' = 1 + \lambda f'', f = S, \theta = 1 + \delta\theta' \quad \text{at} \quad \eta = 0 \tag{2.37a}$$

and

$$f' \to 0, \theta \to 0 \quad \text{as} \quad \eta \to \infty, \tag{2.37b}$$

where the prime denotes differentiation with respect to η. $M = \sqrt{\frac{2\sigma B_0^2 L}{\rho U_0}}$ is the magnetic parameter, $\lambda = N_1 \sqrt{\frac{U_0 \nu}{2L}}$ is the velocity slip parameter, $\delta = D_1 \sqrt{\frac{U_0}{2\nu L}}$ is the thermal slip parameter, $S = \dfrac{V_0}{\sqrt{\dfrac{U_0 \nu}{2L}}} > 0$ (or <0) is the suction (or blowing) parameter, $R = \dfrac{4\sigma^* T_\infty^3}{\kappa k^*}$ is the radiation parameter, and $Pr = \dfrac{\mu c_p}{\kappa}$ is the Prandtl number.

2.3.2 Numerical solutions and analysis of the results

Using shooting method numerical calculations up to the desired level of accuracy were carried out for different values of dimensionless parameters of the problem under consideration for the purpose of illustrating the results graphically. To judge the accuracy of the numerical scheme, comparisons of the present results corresponding to the values of heat transfer coefficient $[-\theta'(0)]$ for $\lambda = 0$, $\delta = 0$ and $M = 0$, $S = 0$ (i.e. in the absence of slip, magnetic field, and suction at the boundary) and also in the absence of thermal radiation are made with the available results of Magyari and Keller [41], Bidin and Nazar [46], El-Aziz [47] and Ishak [49] (for some special cases) and presented in Table 2.1. Results are found to agree well with their results.

In Fig. 2.5(a), velocity profiles are shown for different values of λ. The velocity curves show that the rate of transport decreases with the increasing distance η normal to the sheet. In all cases, the velocity vanishes at some large distance from the sheet at $\eta = 6$. With increasing λ, the stream-wise component of the velocity is found to decrease. When slip occurs, the flow velocity near the sheet is no longer equal to the stretching velocity of the sheet. With an increase in λ, such slip velocity increases and consequently fluid velocity decreases because under the slip condition, the pulling of the stretching sheet can be only partly transmitted to the fluid. With increasing λ, the temperature is found to decrease initially, but after a certain distance from the sheet it increases with λ (Fig. 2.5(b)).

Figure 2.6 depicts the effects of the thermal slip parameter δ on temperature. Initially, the temperature decreases with thermal slip parameter δ, but after a certain distance η normal to the sheet, this feature is smeared out. With an increase of thermal slip parameter δ, less heat is transferred to the fluid from the sheet and so the temperature is found to decrease.

Table 2.1 **Values of $[-\theta'(0)]$ for several values of the Prandtl number in the absence of thermal radiation [1]**

Pr	Magyari and Keller [41]	Bidin and Nazar [46]	El-Aziz [47]	Ishak [49]	Present study with $M = 0$, $S = 0$, $R = 0$ and $\lambda = 0, \delta = 0$
1	0.954782	0.9547	0.954785	0.9548	0.9547
2		1.4714		1.4715	1.4714
3	1.869075	1.8691	1.869074	1.8691	1.8691
5	2.500135		2.500132	2.5001	2.5001

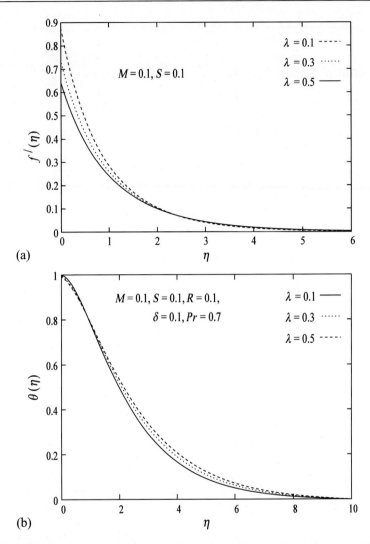

Figure 2.5 Effects of velocity slip parameter λ on (a) velocity and (b) temperature profiles [1].

    Figure 2.7(a) exhibits the nature of the skin-friction coefficient $f''(0)$ with suction/ blowing parameter S for two values of velocity slip ($\lambda = 0.1, 0.3$). It is found that $f''(0)$ increases with S whereas it decreases with higher slip velocity at the boundary. Figure 2.7(b) presents the behavior of heat transfer rate $\theta'(0)$ with suction/blowing parameter S for two values of velocity slip ($\lambda = 0.1, 0.3$). It is very clear that the heat transfer rate increases with blowing but decreases with velocity slip. Figure 2.7(c) displays the nature of heat transfer rate against thermal slip parameter δ for two values of the radiation parameter R. Here, $\theta'(0)$ decreases with thermal slip parameter δ but increases with the radiation parameter R.

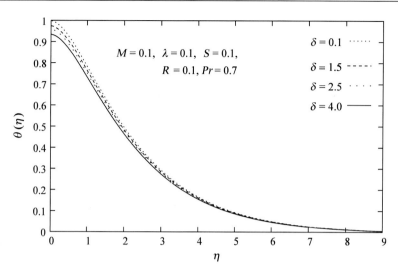

Figure 2.6 Effects of thermal slip parameter δ on temperature profiles [1].

2.4 Flow past an unsteady stretching sheet

In most of the studies related to stretching surface, the flow and temperature fields are considered to be at steady state. However, in some cases, the flow field, heat, and mass transfer can be unsteady because of a sudden stretching of the flat sheet or because of a step change of the temperature of the sheet. When the surface is impulsively stretched with a certain velocity, the inviscid flow is developed instantaneously. However, the flow in the viscous layer near the sheet is developed slowly, and it becomes a fully developed steady flow after a certain instant of time. There have been relatively few papers published for when the stretching force and surface temperature are varying with time. Andersson et al. [52], Dandapat et al. [53], Ali and Magyari [54], and Dandapat et al. [55] studied the problem for unsteady isothermal stretching surface by using a similarity method to transform governing time-dependent boundary layer equations into a set of ordinary differential equations. Elbashbeshy and Bazid [56] have presented similarity solutions of the boundary layer equations that describe the unsteady flow and heat transfer over an unsteady stretching sheet. Sharidan et al. [57] studied the unsteady flow and heat transfer over a stretching sheet in a viscous, incompressible fluid. Liu and Andersson [58] studied thermal characteristics of a liquid film driven by an unsteady stretching surface with prescribed temperature variation of the stretching sheet. Recently, Tsai et al. [59], Chamkha et al. [60], Mukhopadhyay [61–63], and Mukhopadhyay et al. [64] obtained similarity solutions for unsteady flow and heat transfer over a stretching sheet under different conditions. Of late, Bhattacharyya et al. [65] discussed the mass transfer over an unsteady stretching surface in the presence of a first-order chemical reaction.

It is sometimes physically interesting to examine the flow and thermal characteristics of viscous incompressible fluids over a stretching sheet in a porous medium.

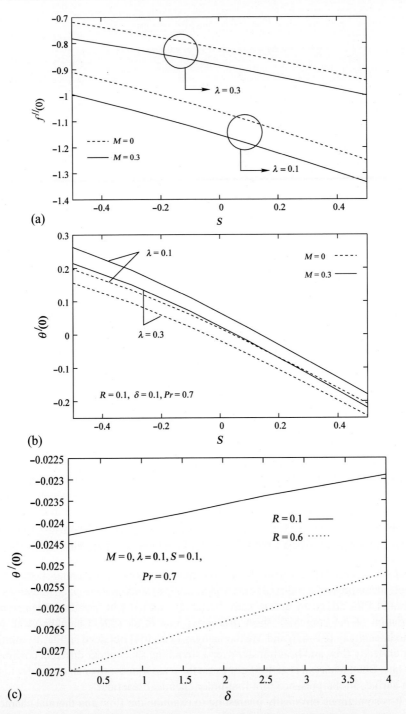

Figure 2.7 Effects of velocity slip parameter λ on (a) skin friction and (b) heat transfer coefficients. (c) Effects of thermal slip parameter δ on heat transfer coefficient [1].

In the physical process of drawing a sheet from a slit of a container, it is tacitly assumed that only the fluid adhered to the sheet is moving but the porous matrix remains fixed to follow the usual assumption of fluid flow in a porous medium. Different models of the porous medium have been formulated, namely, the Darcy, Brinkman, Darcy Brinkman and Forchheimer models. However, the Darcy Brinkman model is widely accepted as most appropriate. Comprehensive reviews of convection through a porous media can be found in Pop and Ingham [66], Ingham and Pop [67], Bejan et al. [68], Ingham et al. [69], Ingham and Pop [70], Vafai [71], and Mukhopadhyay and Layek [72,73]. In this section, we shall present the results obtained for flow past an unsteady stretching sheet and hope that the results obtained will provide useful information for applications.

2.4.1 Mathematical formulation of the problem

We consider laminar boundary-layer flow and heat transfer of viscous incompressible fluid over an unsteady stretching sheet in a porous medium with permeability $k(t) = k_0(1 - at)$, where k_0 is the initial permeability. $k(t)$ is assumed to vary as a linear function of time. We assume that for time $t < 0$, the fluid and heat flows are steady. The unsteady fluid and heat flows start at $t = 0$. The sheet emerges out of a slit at the origin $(x = 0, y = 0)$ and moves with nonuniform velocity $U(x, t) = \frac{cx}{1-at}$, where c, a are positive constants with dimensions (time)$^{-1}$, c is the initial stretching rate, and $\frac{c}{1-at}$ is the effective stretching rate, which is increasing with time. In case of polymer extrusion, the material properties of the extruded sheet may vary with time.

The governing equations of this type of flow are, in the usual notation,

$$\frac{\partial u}{\partial x} + \frac{\partial v}{\partial y} = 0, \tag{2.38}$$

$$\frac{\partial u}{\partial t} + u\frac{\partial u}{\partial x} + v\frac{\partial u}{\partial y} = \nu\frac{\partial^2 u}{\partial y^2} - \frac{\nu}{k}u, \tag{2.39}$$

$$\frac{\partial T}{\partial t} + u\frac{\partial T}{\partial x} + v\frac{\partial T}{\partial y} = \kappa\frac{\partial^2 T}{\partial y^2} + \frac{Q}{\rho c_p}(T - T_\infty), \tag{2.40}$$

when the viscous dissipation term in the energy equation is neglected (as the fluid velocity is low). The linear Darcy term representing distributed body force due to porous media is retained while the nonlinear Forchheimer term is neglected. Here u and v are the components of velocity, respectively, in the x and y directions; μ is the coefficient of fluid viscosity; ρ is the fluid density; ν is the kinematic viscosity of the fluid; T is the temperature; κ is the thermal diffusivity of the fluid; $Q(t) = \frac{Q_0}{1-at}$ is the heat generation ($Q > 0$) or absorption ($Q < 0$) coefficient, which changes with time, Q_0 being the initial value of heat generation/absorption coefficient; and c_p the specific heat.

The appropriate boundary conditions for the problem are given by

$$u = U(x,t), v = 0, T = T_w(x,t) \text{ at } y = 0,$$ (2.41a)

$$u \to 0, T \to T_\infty \text{ as } y \to \infty.$$ (2.41b)

where the temperature of the surface of the sheet is similarly assumed to vary both along the sheet and with time, in accordance with

$T_w(x,t) = T_\infty + \frac{dx^r}{\nu} T_0 (1 - \alpha t)^{-s}$ (Liu and Andersson [58]), where T_0 is the fixed slit temperature at $x = 0$ (except for $r = 0$), d is the constant of proportionality assumed to be positive with dimension (length $^{2-r}$ time $^{-1}$), and T_∞ is the constant free stream temperature. The power indices r, s enable us to examine a variety of different temperature variations. With $r > 0$, the sheet temperature decreases as x^r with the distance from the slit. Similarly, for $s > 0$, the sheet temperature at a fixed location x is reduced with time in proportion to $(1 - \alpha t)^{-s}$. The expressions for $U(x,t), T_w(x,t), k(t), Q(t)$ are valid for time $t < \alpha^{-1}$. These forms are chosen in order to get a new similarity transformation, which transforms the governing partial differential equations into a set of ordinary differential equations, thereby facilitating the exploration of the effects of the controlling parameters.

We now introduce the following relations for u, v and θ as

$$u = \frac{\partial \psi}{\partial y}, v = -\frac{\partial \psi}{\partial x} \text{ and } \theta = \frac{T - T_\infty}{T_w - T_\infty}$$ (2.42)

where ψ is the stream function.

We introduce

$$\eta = \sqrt{\frac{c}{\nu(1 - \alpha t)}} y, \psi = \sqrt{\frac{\nu c}{(1 - \alpha t)}} x f(\eta), T = T_\infty + T_0 \left[\frac{dx^r}{\nu} \right] (1 - \alpha t)^{-s} \theta(\eta).$$

With the help of the above relations, the governing equations finally reduce to

$$M \left(\frac{\eta}{2} f'' + f' \right) + f'^2 - ff'' = f''' - k_1 f,$$ (2.43)

$$\frac{M}{2} \eta \theta' + sM\theta + rf'\theta - f\theta' = \frac{1}{Pr} \theta'' + \lambda \theta,$$ (2.44)

where $M = \frac{\alpha}{c}$ is the unsteadiness parameter, $k_1 = \frac{\nu}{k_0 c}$ is the permeability parameter,

$Pr = \frac{\nu}{\kappa}$ is the Prandtl number, and $\lambda = \frac{Q_0}{\rho c c_p}$ is the heat source or sink parameter.

The boundary conditions then become

$$f' = 1, f = 0, \theta = 1 \text{ at } \eta = 0$$ (2.45a)

and

$$f' \to 0, \theta \to 0 \text{ as } \eta \to \infty.$$ (2.45b)

2.4.2 Numerical solutions and discussion of the results

In order to validate the method used in this study and to judge the accuracy of the present analysis, comparison is made with available results of Sharidan et al. [57] and Chamkha et al. [60] in Table 2.2, and it is found that the results agree well.

With both r and s equal to zero, the surface temperature defined earlier simplifies to $T_w = T_\infty - T_0 \frac{d}{v}$, that is, a constant sheet temperature. With $r = s = 0$ and also with $\lambda = 0$ in equation (2.44), the trivial solution $\theta(\eta) = 1$ is readily found to satisfy the boundary conditions in equations (2.45a) and (2.45b). This solution of the isothermal sheet problem was also reported by Andersson et al. [52]. Moreover, if the temperature $\theta(\eta)$ in the thermal energy equation (2.44) is replaced by the velocity $f'(\eta)$, it can easily be demonstrated that the thermal energy equation (2.44) becomes identical with the momentum equation (2.43) in the absence of porous media if, and only if, the Prandtl number Pr is set equal to unity and the power indices $r = s = 1, \lambda = 0$. This observation implies that the solution $f'(\eta)$ of the hydrodynamic problem (in case of nonporous media) also solves the thermal energy problem for this particular parameter combination (Liu and Andersson [56]).

Figure 2.8(a) exhibits the velocity profiles for several values of unsteadiness parameter M. It is seen that the velocity along the sheet decreases with the increase of unsteadiness parameter M, and this implies an accompanying reduction of the thickness of the momentum boundary layer. With increasing unsteadiness parameter M, the

Table 2.2 The values of $f''(0)$ for various values of unsteadiness parameter M in nonporous medium

M	Sharidan et al. [57]	Chamakha et al. [60]	Present study
0.8	−1.261042	−1.261512	−1.261479
1.2	−1.377722	−1.378052	−1.377850

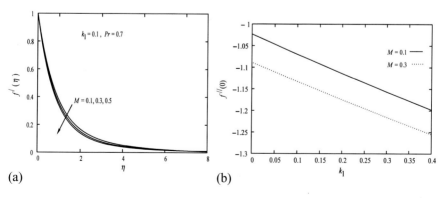

(a) (b)

Figure 2.8 Effects of unsteadiness parameter M on (a) velocity profiles and (b) skin-friction coefficient.

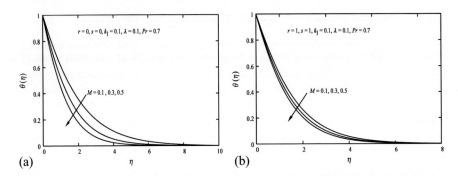

Figure 2.9 Temperature profiles for prescribed (a) constant (b) variable surface temperature.

skin-friction coefficient $f''(0)$ decreases [see Fig. 2.8(b)]. The skin-friction coefficient decreases with permeability parameter k_1 [see Fig. 2.8(b)].

Figure 2.9(a) and (b) represent the effects of the unsteadiness parameter on the temperature distribution at prescribed constant surface temperature and variable surface temperature, respectively. From both the figures, it is noticed that the temperature at a particular point is found to decrease significantly with increasing unsteadiness parameter. But the decrease in temperature $\theta(\eta)$ is more pronounced in case of constant sheet temperature (Fig. 2.9(a)). Rate of heat transfer increases with increasing M: As the unsteadiness parameter M increases, less heat is transferred from the sheet to the fluid; hence, the temperature $\theta(\eta)$ decreases (Fig. 2.9(a) and (b)). Because the fluid flow is caused solely by the stretching of the sheet, and the sheet surface temperature is higher than free stream temperature, the velocity and temperature decrease with increasing η. It is important to note that the rate of cooling is much faster for higher values of the unsteadiness parameter whereas it may take a longer time for cooling during steady flows.

The rate of heat transfer $\theta'(0)$ increases with increasing heat source/sink parameter λ and also with the parameter r (Fig. 2.10).

2.5 Flow past a curved stretching sheet

When the external flow, that is, the flow immediately adjacent to the boundary layer region is curved, the development of the boundary layer is greatly affected by the normal pressure gradient toward the contour of curvature and gives rise to "crosswise flow." Such problems are very complicated because of the interaction effects of boundary layer flowing over this very sharp curved surface and due to the sharp curvature itself. The flow due to a curved stretching surface may have applications in a stretch-forming machine with curving jaws (see Sajid et al. [74,75]).

In most of the studies mentioned above, the stretching surface is flat and has no curvature. In these problems, mathematical modeling is carried out using Cartesian

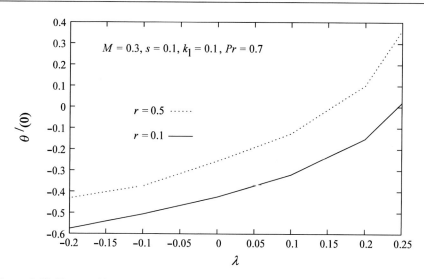

Figure 2.10 Nature of heat transfer coefficient with spatial index r.

coordinates. But the flow past a curved stretching surface is of a different kind. To our knowledge, there is scarcity in the literature regarding the stretching flow over a curved surface. Mathematical modeling in case of a curved stretching surface with uniform curvature was first handled by Sajid et al. [74]. In discussing the two-dimensional flow for a stretching sheet having uniform curvature, they obtained the governing equations using a curvilinear coordinate system. They pointed out that the pressure inside the boundary layer is not zero as it was for a flat stretching sheet. Sajid et al. [75] extended the stretching flow of a micropolar fluid for a curved sheet. Abbas et al. [76] studied the laminar flow and heat transfer analysis of an electrically conducting viscous fluid over a curved stretching surface with constant as well as variable surface temperature. Hence, we shall present here in this section the results for flow past a curved stretching sheet, and hope that the results presented here provide useful information for applications.

2.5.1 Formulation of the problem

Consider a two-dimensional boundary layer flow of an incompressible viscous fluid over a curved stretching surface coiled in a circle of radius R. Under boundary layer approximations, the governing equations for the flow in curvilinear coordinates, are given by

$$\frac{\partial}{\partial r}\{(R+r)v\} + R\frac{\partial u}{\partial s} = 0, \tag{2.46}$$

$$\frac{u^2}{r+R} = \frac{1}{\rho}\frac{\partial p}{\partial r}, \tag{2.47}$$

$$v\frac{\partial u}{\partial r}+\frac{Ru}{r+R}\frac{\partial u}{\partial s}+\frac{uv}{r+R}=-\frac{1}{\rho}\frac{R}{r+R}\frac{\partial p}{\partial s}+v\left(\frac{\partial^2 u}{\partial r^2}+\frac{1}{r+R}\frac{\partial u}{\partial r}-\frac{u}{(r+R)^2}\right). \qquad (2.48)$$

Here, $v=\frac{\mu}{\rho}$ is the kinematic viscosity, μ is the coefficient of viscosity, ρ is the fluid density, r is the radial coordinate, s is the arc length coordinate, v and u are the velocity components in the r and s directions, respectively, and R is the distance of the sheet from the origin; large values of R correspond to small curvature.

The energy equation is given by

$$\rho c_p\left[r\frac{\partial T}{\partial r}+\frac{R}{r+R}u\frac{\partial T}{\partial s}\right]=\kappa\left[\frac{\partial^2 T}{\partial r^2}+\frac{1}{r+R}\frac{\partial T}{\partial r}\right], \qquad (2.49)$$

where T is the temperature and κ is the thermal conductivity.

The appropriate boundary conditions are

$$u=as, v=-v_w, T=T_w \quad \text{at} \quad r=0, \qquad (2.50a)$$

$$u\to 0, \frac{\partial u}{\partial r}\to 0, T\to T_\infty \quad \text{as} \quad r\to\infty, \qquad (2.50b)$$

where $a>0$ is the stretching constant and $v_w=\frac{R}{r+R}v_0$ is the variable velocity at the wall and corresponds to suction when $v_0>0$, v_0 being the initial value.

Let us introduce the similarity transformations

$$u=asf'(\eta), v=-\frac{R}{r+R}\sqrt{av}f(\eta), p=\rho a^2 s^2 P(\eta),$$

$$\eta=\sqrt{\frac{a}{v}}r, \theta(\eta)=\frac{T-T_w}{T_w-T_\infty}. \qquad (2.51)$$

Using these transformations, equation (2.46) is automatically satisfied and equations (2.47) and (2.48) become

$$\frac{\partial P}{\partial\eta}=\frac{f'^2}{\eta+K}, \qquad (2.52)$$

$$\frac{2K}{\eta+K}P=f'''+\frac{1}{\eta+K}f''-\frac{1}{(\eta+K)^2}f'-\frac{K}{\eta+K}f'^2+\frac{K}{\eta+K}ff''+\frac{K}{(\eta+K)^2}ff'. \qquad (2.53)$$

Eliminating pressure P, we finally get

$$f^{iv}+\frac{2}{\eta+K}f'''-\frac{1}{(\eta+K)^2}f''+\frac{1}{(\eta+K)^3}f'-\frac{K}{(\eta+K)}\left(f'f''-ff'''\right)$$

$$-\frac{K}{(\eta+K)^2}\left(f'^2-ff''\right)-\frac{K}{(\eta+K)^3}ff'=0. \qquad (2.54)$$

The energy equation finally takes the form

$$\theta^{//} + \frac{1}{(\eta + K)}\theta^{/} + Pr\frac{K}{(\eta + K)}f\theta^{/} = 0, \tag{2.55}$$

$$f(0) = S, f^{/}(0) = 1, \theta(0) = 1, \tag{2.56a}$$

$$f^{/}(\infty) = 0, f^{//}(\infty) = 0, \theta(\infty) = 0. \tag{2.56b}$$

Here $K = R\sqrt{\frac{a}{\nu}}$ is the curvature parameter, $S = \frac{v_0}{\sqrt{a\nu}}(>0)$ is the suction parameter.

2.5.2 Numerical solutions and discussion of the results

The highly nonlinear equations (2.54) and (2.55) subject to the boundary conditions (2.56a,b) are solved numerically by the shooting method. The quantities of main interest are the velocity and temperature distributions. From Fig. 2.11(a), it is noted that the velocity (for the horizontal component of velocity u) and boundary layer thickness increase with the decreasing values of the curvature parameter K (i.e., increase in dimensionless curvature). Temperature decreases with the increase in the values of curvature parameter K (Fig. 2.11(b)). Figure 2.12(a) and (b) show the effects of suction parameter S on velocity and temperature fields, respectively. Both velocity and temperature are found to decrease with the increasing suction. It is also a well-established phenomenon that suction can be applied to control the boundary layer thickness and can also be used as a means of cooling in heat transfer problems.

2.6 Stagnation point flow of a non-newtonian fluid over a stretching sheet

Stagnation-point flow, which is a classical problem in fluid mechanics, has enormous applications in manufacturing industries such as polymer extraction, hot rolling and wire drawing, paper production, glass fiber production, metal spinning, plastic film drawing, and many others. The stagnation region encounters the highest pressure and the highest heat transfer. The classical viscous flows near a stagnation point toward a rigid horizontal plane and a rigid horizontal axisymmetric surface are very popular as they admit exact solutions for the Navier–Stokes equations. Hiemenz and Homann analyzed these, and these are described in detail in the textbook *Boundary Layer Theory* by Schlichting [77]. Hiemenz [78] was the first person who discovered that the stagnation point flow can be analyzed exactly by the Navier–Stokes equations, and for the two-dimensional case, he presented the velocity distribution. Later, the corresponding temperature distribution was reported by Goldstein [79]. Since then, many investigators have considered various aspects of the problem.

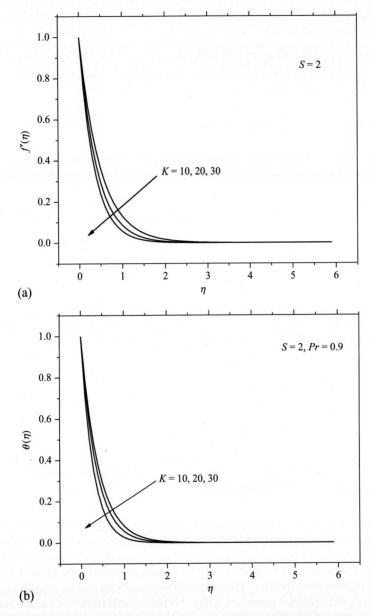

Figure 2.11 Effects of curvature parameter K on (a) velocity and (b) temperature profiles.

This plane stagnation point flow is referred to as the Hiemenz flow. Gesten et al. [80] extended the problem and obtained the results for three-dimensional stagnation point flow. Liao [81] found an explicit, totally analytic, uniformly valid solution of the two-dimensional laminar viscous flow over a semi-infinite plate via an analytic technique, namely, the HAM.

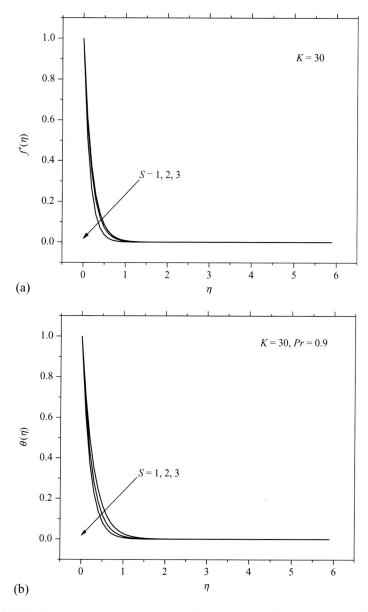

Figure 2.12 Effects of suction parameter S on (a) velocity and (b) temperature profiles.

Stagnation-point flow is a significant topic in fluid mechanics in the sense that stagnation points appear in virtually all flow fields of science and engineering. Near the stagnation region, fluid motion exists on all solid bodies moving in a fluid. In some situations, flow is stagnated by a solid wall, whereas in others a free stagnation point or a line exists in the interior of the fluid domain. These flows were characterized as inviscid or viscous, steady or unsteady, two-dimensional or three-dimensional, symmetric or asymmetric, normal or oblique, homogeneous or two-fluid, and forward or reverse after the investigation of Weidman and Putkaradze [82]. A number of authors, such as Stuart [83], Tamada [84], Takemitsu and Matunobu [85], Dorrepaal [86,87], Labropulu et al. [88], and Tilley and Weidman [89], investigated the two-dimensional stagnation point flow impinging obliquely on a fixed plane wall. Two-dimensional stagnation-point flow of a power-law fluid toward a rigid surface was investigated by Kapur and Srivastava [90]. Later, Maiti [91] and Koneru and Manohar [92] extended the same problem to the axisymmetric case. The two-dimensional orthogonal stagnation-point flow of an incompressible electrically conducting power-law fluid toward a rigid surface in presence of a uniform transverse magnetic field was investigated by Sapunkov [93]. Djukic [94] studied the hydromagnetic Hiemenz flow of a power-law fluid toward a rigid plate.

Later, Mahapatra and Gupta [95], Nazar et al. [96], Lok et al. [97], Reza and Gupta [98], and Mahapatra et al. [99] considered the following problem:

$$f''' + ff'' - \left(f'\right)^2 + \frac{a^2}{c^2} = 0, \tag{2.57}$$

with the boundary conditions

$$f(0) = 0, f'(0) = 1, \ \lim_{\eta \to \infty} f'(\eta) = \frac{a}{c}. \tag{2.58}$$

It describes the flow toward the orthogonal stagnation point on a horizontal stretching sheet. The parameter a/c is defined by $u_e(x)/u_w(x) = a/c$, where $u_e(x) = ax$ is the velocity of the flow outside the boundary layer (inviscid flow) and $u_w(x) = cx$ is the velocity of the stretching sheet, respectively, a and c being positive constants. Lok et al. [97] pointed out that the flow has a boundary layer behavior when $a/c > 1$ and it has an inverted boundary layer structure when $a/c < 1$. The existence, uniqueness, and stability of a monotonic physically meaningful solution for $a/c > 0$ was established by Paullet and Weidman [100]. If $a/c > 1$, it is found that this is the only solution, whereas in case of $0 < a/c < 1$ there may exist two solutions, only one of which is monotonic. In the nonorthogonal stagnation-point flow, the self-similar flow is governed by a system of three equations, namely

$$f''' + ff'' - \left(f'\right)^2 + \frac{a^2}{c^2} = 0, \tag{2.59}$$

$$G''' + f G'' - f' G' + \lambda H + \text{const} = 0, \tag{2.60}$$

$$H'' + PrH'f = 0, \tag{2.61}$$

with the boundary conditions

$$f(0) = 0, f'(0) = 1, \lim_{\eta \to \infty} f'(\eta) = \frac{a}{c}, G(0) = G'(0) = 0, H(0)$$

$$= 1, G''(0) = \gamma, \lim_{\eta \to \infty} H(\eta) = 0. \tag{2.62}$$

The nonorthogonal stagnation point flow shown above is considered in Lok et al. [101]. For the related mixed convection problems, see [102–105]. Several authors (see, for instance, [106–108]) have considered the hydromagnetic stagnation point flow over a stretching sheet. Later, several researchers (see, for instance, [109–111]) investigated the case of stagnation point flow in a porous medium. Furthermore, to allow for suction at the surface, Kechil and Hashim [111] considered the modified boundary condition (see Vajravelu and Van Giorder [112]).

In most of the investigations mentioned above, the researchers restricted their investigations to Newtonian fluid. The flows of non-Newtonian fluids are very important because of their industrial and technological applications. To obtain a thorough cognition of non-Newtonian fluids and their various applications, it is necessary to study their flow behaviors (Mukhopadhyay and Vajravelu [113]). The governing equations of non-Newtonian fluids are highly nonlinear and much more complicated than that of Newtonian fluids. Because of the complexity of these fluids, there is not a single constitutive equation that exhibits all properties of non-Newtonian fluids. As a result, a number of non-Newtonian fluid models have been proposed. The Casson fluid model is a non-Newtonian fluid model that exhibits yield stress. If a shear stress less than the yield stress is applied to the fluid, then it behaves like a solid, whereas if a shear stress greater than the yield stress is applied, it starts to move. Examples of Casson fluid include jelly, tomato sauce, honey, soup, concentrated fruit juices, etc. Human blood can also be treated as a Casson fluid.

Motivated by these studies mentioned above, we shall present the numerical solutions for boundary layer flow of a Casson fluid past a nonlinearly stretching sheet near a stagnation point. So this problem can be considered as an extension of the work of Cortell [114] who investigated the flow of Newtonian fluid and heat transfer past a nonlinearly stretching sheet. Actually in this problem, the work of Mahapatra et al. [115] is also extended by considering non-Newtonian fluid obeying the Casson fluid model. The existence of dual solutions to nonlinear equations representing physical problems is important from theoretical as well as practical points of view. This gives the fundamental mechanism of the flow problem under investigation. We shall present here the dual solutions and its range of existence for stagnation-point flow of a non-Newtonian fluid over a stretching sheet.

2.6.1 Formulation of the mathematical problem

The rheological equation of state for an isotropic and incompressible flow of a Casson fluid is given by

$$\tau_{ij} = \begin{cases} 2\left(\mu_B + p_y/\sqrt{2\pi}\right)e_{ij}, \pi > \pi_c, \\ 2\left(\mu_B + p_y/\sqrt{2\pi_c}\right)e_{ij}, \pi < \pi_c. \end{cases}$$

Here $\pi = e_{ij}e_{ij}$ and e_{ij} is the (i,j)-th component of the deformation rate, π is the product of the component of deformation rate with itself, π_c is a critical value of this product based on the non-Newtonian model, μ_B is plastic dynamic viscosity of the non-Newtonian fluid, and p_y is the yield stress of the fluid.

Consider a flow of an incompressible viscous fluid past a flat sheet coinciding with the plane $y=0$. The fluid flow is confined to $y>0$. Two equal and opposite forces are applied along the x-axis so that the wall is stretched keeping the origin fixed. The equations of motion in usual notation for steady stagnation point flow using the boundary layer approximation for Casson fluid are

$$\frac{\partial u}{\partial x} + \frac{\partial v}{\partial y} = 0, \tag{2.63}$$

$$u\frac{\partial u}{\partial x} + v\frac{\partial u}{\partial y} = U\frac{dU}{dx} + \nu\left(1 + \frac{1}{\beta}\right)\frac{\partial^2 u}{\partial y^2}, \tag{2.64}$$

where u, v are the velocity components in the x and y directions respectively, ν is the kinematic viscosity of the fluid, $\beta = \mu_B\sqrt{2\pi_c}/p_y$ is the parameter of the Casson fluid, $U = U(x) = a\,x^n$ is the free stream velocity, and a is a parameter related to surface stretching speed.

The boundary conditions are

$$u = cx^n, v = 0 \quad \text{at} \quad y = 0, \tag{2.65a}$$

$$u \to U(x) = ax^n \quad \text{as} \quad y \to \infty \tag{2.65b}$$

where $c > 0$ is the stretching rate and n is an exponent.

We are interested to find the self-similar form of the above governing equations. The similarity variable η and the velocity components u, v are given below

$$\eta = \sqrt{\frac{c(n+1)}{2\nu}}\,yx^{\frac{n-1}{2}}, u = cx^n f'(\eta),$$

$$v = -\sqrt{\frac{c\nu(n+1)}{2}}\left[f(\eta) + \left(\frac{n-1}{n+1}\right)\eta f'(\eta)\right]x^{\frac{n-1}{2}}. \tag{2.66}$$

These self-similar structure functions (2.66) reduce the equations (2.63) and (2.64) into an ordinary differential equation as

$$\left(1 + \frac{1}{\beta}\right)f'''(\eta) + f(\eta)f''(\eta) + \left(\frac{2n}{n+1}\right)\left[\alpha^2 - f'(\eta)^2\right] = 0 \tag{2.67}$$

where $\alpha(=a/c)$ is the velocity ratio parameter. The boundary conditions reduce to

$$f(0) = 0, f'(0) = 1, \lim_{\eta \to \infty} f'(\eta) = \alpha. \tag{2.68}$$

The standard shooting technique with fourth-order Runge-Kutta method and Newton's method are used for finding approximate solutions of the above equation (2.67).

2.6.2   Results and discussion

In order to validate the numerical method used in this study and to judge the accuracy of the present analysis, comparisons with available results are presented here. The results for the Newtonian case, that is, for $\beta \to \infty$, agree well with the results of other authors, which can be found from Table 2.3.

Therefore, we believe that our results are accurate and reliable and, hence, this gives confidence to study further the problem for Casson fluid.

In order to analyze the results, numerical computations have been carried out using the method stated in the previous section for various values of the velocity ratio parameter α, nonlinear stretching parameter n, and the Casson parameter β.

The results obtained for the flow characteristics reveal many interesting behaviors that justify further study of the equations related to non-Newtonian fluid phenomena.

Figure 2.13 exhibits the nature of the skin friction coefficient with the velocity ratio parameter (α) for various values of the nonlinearly stretching parameter n for Casson fluid ($\beta = 2$). From this figure, it is observed that dual solutions exist for all values of n considered in this study, and the range of the dual solutions increases with increasing values of n (Fig. 2.13). Skin friction decreases with increasing values of n. It is observed that the range of dual solutions is slightly larger for a Casson fluid than a Newtonian fluid (Fig. 2.14). Shear stress at the wall is negative here. Physically, a negative sign implies that the surface exerts a dragging force on the fluid, and a positive sign implies the opposite. It is found that dual solutions for Newtonian fluid exist in the following ranges of α:

(i) for nonlinearly stretching sheet ($n=0.5$), $0<\alpha<0.099446$
(ii) for linearly stretching sheet ($n=1.0$), $0<\alpha<0.169056$.

On the other hand, for a non-Newtonian Casson fluid, the following ranges of α are found:

(i) for nonlinearly stretching sheet ($n=0.5$) and with $\beta=2.0$, $0<\alpha<0.099643$
(ii) for linearly stretching sheet ($n=1.0$) and with $\beta=2.0$, $0<\alpha<0.169061$.

Figure 2.15(a) shows that fluid velocity decreases with the increasing values of n. The effects of the nonlinear stretching parameter n are significant when n is low. For a fixed value of n, fluid velocity is higher for a non-Newtonian fluid than for a Newtonian fluid (Fig. 2.15(b)). Fluid velocity is found to decrease with increasing values of β for $n=0.5$ (Fig. 2.15(b)). The momentum boundary layer thickness decreases with increasing Casson parameter β.

Table 2.3 Comparison and new results of wall shear stress $-f''(0)$ when $\alpha = 0$ for non-Newtonian fluid

n	Other studies		Present study				
	Cortell [114]	Mahapatra et al. [115]	$\beta = \infty$, Newtonian Case	$\beta = 0.5$	$\beta = 1.0$	$\beta = 2.0$	$\beta = 5.0$
0.0	0.627547	0.627554	0.627555	0.362319	0.443748	0.512396	0.527877
0.2	0.766758	0.766837	0.766837	0.442734	0.542236	0.626120	0.700023
0.5	0.889477	0.889544	0.889544	0.513578	0.629003	0.726309	0.812039
0.75	0.953786	0.953957	0.953957	0.550767	0.674549	0.778902	0.870839
1.0	1.0	1.0	1.0	0.577351	0.707107	0.816497	0.912871
1.5	1.061587	1.061601	1.061601	0.612916	0.750665	0.866794	0.969105
3.0	1.148588	1.148590	1.148593	0.663141	0.812178	0.937822	1.048517
5.0	—	1.194486	1.194487	0.689638	0.844630	0.975295	1.090413
7.0	1.216847	1.216851	1.216851	0.702549	0.860443	0.993554	1.110827
10.0	1.234875	1.234870	1.234874	0.712955	0.873188	1.008271	1.127281
20.0	1.257418	1.257420	1.257424	0.725974	0.889133	1.026682	1.147865
100.0	1.276768	1.276771	1.276773	0.737146	0.902815	1.042481	1.165530

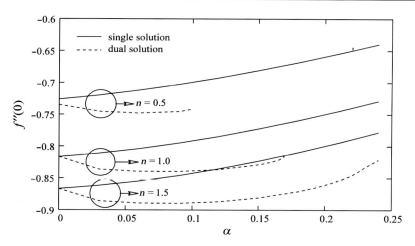

Figure 2.13 Skin friction coefficient against α for a non-Newtonian Casson fluid with $\beta = 2$ for various values of n.

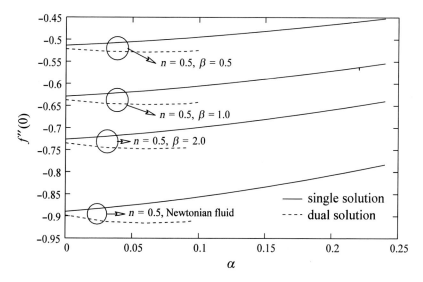

Figure 2.14 Skin friction co-efficient against α for a Newtonian and non-Newtonian Casson fluid with different values of β.

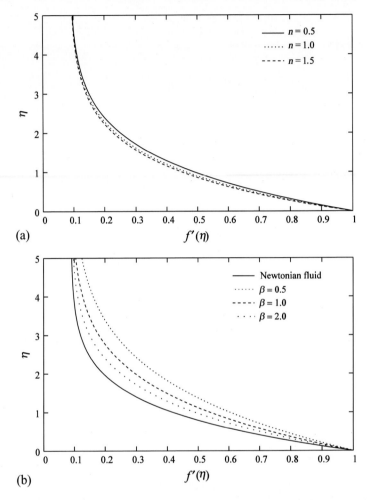

Figure 2.15 Velocity profiles for different values of (a) nonlinearly stretching parameter n with $\alpha = 0.09$, $\beta = 2.0$ and (b) Casson parameter β for nonlinearly stretching sheet ($n = 0.5$) with $\alpha = 0.09$.

References

[1] Mukhopadhyay S. Slip effects on MHD boundary layer flow over an exponenttially stretching sheet with suction/blowing and thermal radiation. Ain Shams Eng J 2013;4:485–91.

[2] Mukhopadhyay S, Gorla RSR. Effects of partial slip on boundary layer flow past a permeable exponential stretching sheet in presence of thermal radiation. Heat Mass Transfer 2012;48:1773–81.

[3] Gupta PS, Gupta AS. Heat and mass transfer on a stretching sheet with suction or blowing. Can J Chem Eng 1977;55:744–6.

[4] Prasad KV, Vajravelu K, Datti PS. Mixed convection heat transfer over a non-linear stretching surface with variable fluid properties. Int J Non-Linear Mech 2010;45:320–30.

[5] Zheng L, Niu J, Zhang X, Ma L. Dual solutions for flow and radiative heat transfer of a micropolar fluid over stretching/shrinking sheet. Int J Heat Mass Tran 2012;55:7577–86.

[6] Pal D. Heat and mass transfer in stagnation-point flow towards a stretching surface in the presence of buoyancy force and thermal radiation. Meccanica 2009;44:145–58.

[7] Vajravelu K, Prasad KV, Chiu-On Ng. Unsteady flow and heat transfer in a thin film of Ostwald–de Waele liquid over a stretching surface. Comm Nonlinear Sci Numer Simulat 2012;17:4163–73.

[8] Mukhopadhyay S, Layek GC, Gorla RSR. MHD combined convective flow and heat transfer past a porous stretching surface. Int J Fluid Mech Res 2007;34:244–57.

[9] Ray Mahapatra T, Dholey S, Gupta AS. Oblique stagnation-point flow of an incompressible visco-elastic fluid towards a stretching surface. Int J Non-Linear Mech 2007;42:484–99.

[10] Crane LJ. Flow past a stretching plate. Z Angew Math Phys 1970;21:645–7.

[11] Carragher P, Crane LJ. Heat transfer on a continuous stretching sheet. Z Angew Math Mech 1982;62:564.

[12] Chen CK, Char MI. Heat transfer of a continuous stretching surface with suction or blowing. J Math Anal Appl 1988;135:568–80.

[13] Datta BK, Roy P, Gupta AS. Temperature field in the flow over a stretching sheet with uniform heat flux. Int Comm Heat Mass Tran 1985;12:89–94.

[14] Vajravelu K. Convection heat transfer at a stretching sheet with suction or blowing. J Math Anal Appl 1994;188(3):1002–11.

[15] Mukhopadhyay S, Mandal IC, Gorla RSR. Effects of thermal stratification on flow and heat transfer past a porous vertical stretching surface. Heat Mass Transfer 2012;48:915–21.

[16] Ishak A, Nazar R, Pop I. Hydromagnetic flow and heat transfer adjacent to a stretching vertical sheet. Heat Mass Transfer 2008;44:921–7.

[17] Mukhopadhyay S, Layek GC, Samad SkA. Study of MHD boundary layer flow over a heated stretching sheet with variable viscosity. Int J Heat Mass Tran 2005;48:4460–6.

[18] Batchelor GK. An introduction to fluid dynamics. London: Cambridge University Press; 1967, 597.

[19] Mukhopadhyay S, Layek GC. Effects of thermal radiation and variable fluid viscosity on free convective flow and heat transfer past a porous stretching surface. Int J Heat Mass Tran 2008;51:2167–78.

[20] Saikrishnan P, Roy S. Non-uniform slot injection (suction) into water boundary layers over (i) a cylinder and (ii) a sphere. Int J Eng Sci 2003;41:1351–65.

[21] Bird RB, Stewart WE, Lightfoot EN. Transport phenomena. New York: John Wiley and Sons; 1960.

[22] Mukhopadhyay S. Analysis of boundary layer flow over a porous nonlinearly stretching sheet with partial slip at the boundary. Alex Eng J 2013;52:563–9.

[23] Ali ME. On thermal boundary layer on a power law stretched surface with suction or injection. Int J Heat Fluid Flow 1995;16:280–90.

[24] Vajravelu K. Viscous flow over a nonlinearly stretching sheet. Appl Math Comput 2001;124:281–8.

[25] Vajravelu K, Cannon JR. Fluid flow over a nonlinear stretching sheet. Appl Math Comput 2006;181:609–18.

[26] Bataller RC. Similarity solutions for flow and heat transfer of a quiescent fluid over a nonlinearly stretching surface. J Mater Process Technol 2008;203:176–83.

[27] Cortell R. Effects of viscous dissipation and radiation on the thermal boundary layer over a non-linearly stretching sheet. Phys Lett A 2008;372:631–6.

[28] Akyildiz T, Siginer DA, Vajravelu K, Cannon JR, Van Gorder RA. Similarity solutions of the boundary layer equations for a nonlinearly stretching sheet. Math Meth Appl Sci 2010;33:601–6.

[29] Van Gorder RA, Vajravelu K. A note on flow geometries and the similarity solutions of the boundary layer equations for a nonlinearly stretching sheet. Arch Appl Mech 2009; http://dx.doi.org/10.1007/s00419-009-0370-6.

[30] Yoshimura A, Prudhomme RK. Wall slip corrections for Couette and parallel disc viscometers. J Rheol 1998;32:53–67.

[31] Navier CLMH. Mémoire sur les lois du movement des fluids. Mém Acad R Sci Inst France 1823;6:389–416.

[32] Mukhopadhyay S, Golam Arif M, Wazed Ali Pk M. Effects of partial slip on chemically reactive solute transfer in boundary layer flow over an exponentially stretching sheet with suction/blowing. J Appl Mech Tech Phys 2013;54(6).

[33] Wang CY. Flow due to a stretching boundary with partial slip—an exact solution of the Navier-Stokes equations. Chem Eng Sci 2002;57:3745–7.

[34] Andersson HI. Slip flow past a stretching surface. Acta Mech 2002;158:121–5.

[35] Ariel PD, Hayat T, Asghar S. The flow of an elastico-viscous fluid past a stretching sheet with partial slip. Acta Mech 2006;187:29–35.

[36] Ariel PD. Two dimensional stagnation point flow of an elastico-viscous fluid with partial slip. Z Angew Math Mech 2008;88:320–4.

[37] Abbas Z, Wang Y, Hayat T, Oberlack M. Slip effects and heat transfer analysis in a viscous fluid over an oscillatory stretching surface. Int J Numer Meth Fluids 2009;59:443–58.

[38] Mukhopadhyay S, Andersson HI. Effects of slip and heat transfer analysis of flow over an unsteady stretching surface. Heat Mass Transfer 2009;45:1447–52.

[39] Mukhopadhyay S. Effects of slip on unsteady mixed convective flow and heat transfer past a porous stretching surface. Nucl Eng Des 2011;241:2660–5.

[40] Gad-el-Hak M. The fluid mechanics of microdevices: the Freeman Scholar Lecture. ASME J Fluids Eng 1999;199:5–33.

[41] Magyari E, Keller B. Heat and mass transfer in the boundary layers on an exponentially stretching continuous surface. J Phys D Appl Phys 1999;32:577–85.

[42] Elbashbeshy EMA. Heat transfer over an exponentially stretching continuous surface with suction. Arch Mech 2001;53:643–51.

[43] Khan SK. Boundary layer viscoelastic fluid flow over an exponentially stretching sheet. Int J Appl Mech Eng 2006;11:321–35.

[44] Sanjayanand E, Khan SK. On heat and mass transfer in a viscoelastic boundary layer flow over an exponentially stretching sheet. Int J Therm Sci 2006;45:819–28.

[45] Sajid M, Hayat T. Influence of thermal radiation on the boundary layer flow due to an exponentially stretching sheet. Int Comm Heat Mass Tran 2008;35:347–56.

[46] Bidin B, Nazar R. Numerical solution of the boundary layer flow over an exponentially stretching sheet with thermal radiation. Eur J Sci Res 2009;33(4):710–7.

[47] El-Aziz MA. Viscous dissipation effect on mixed convection flow of a micropolar fluid over an exponentially stretching sheet. Can J Phys 2009;87:359–68.

[48] Pal D. Mixed convection heat transfer in the boundary layers on an exponentially stretching surface with magnetic field. Appl Math Comput 2010;217:2356–69.

[49] Ishak A. MHD boundary layer flow due to an exponentially stretching sheet with radiation effect. Sains Malaysiana 2011;40:391–5.

[50] Bhattacharyya K, Mukhopadhyay S, Layek GC. Slip effects on an unsteady boundary layer stagnation-point flow and heat transfer towards a stretching sheet. Chin Phys Lett 2011;28(9):094702.

[51] Brewster MQ. Thermal radiative transfer properties. Hoboken, NJ: Wiley; 1972.

[52] Andersson HI, Aarseth JB, Dandapat BS. Heat transfer in a liquid film on an unsteady stretching surface. Int J Heat Mass Tran 2000;43:69–74.

[53] Dandapat BS, Santra B, Andersson HI. Thermocapillarity in a liquid film on an unsteady stretching surface. Int J Heat Mass Tran 2003;46:3009–15.

[54] Ali ME, Magyari E. Unsteady fluid and heat flow induced by a submerged stretching surface while its steady motion is slowed down gradually. Int J Heat Mass Tran 2007;50:188–95.

[55] Dandapat BS, Santra B, Vajravelu K. The effects of variable fluid properties and thermo-capillarity on the flow of a thin film on an unsteady stretching sheet. Int J Heat Mass Tran 2007;50:991–6.

[56] Elbashbeshy EMA, Bazid MAA. Heat transfer over an unsteady stretching surface. Heat Mass Transfer 2004;41:1–4.

[57] Sharidan S, Mahmood T, Pop I. Similarity solutions for the unsteady boundary layer flow and heat transfer due to a stretching sheet. Int J Appl Mech Eng 2006;11(3): 647–54.

[58] Liu IC, Andersson HI. Heat transfer in a liquid film on an unsteady stretching sheet. Int J Therm Sci 2008;47:766–72.

[59] Tsai R, Huang KH, Huang JS. Flow and heat transfer over an unsteady stretching surface with a non-uniform heat source. Int Comm Heat Mass Tran 2008;35:1340–3.

[60] Chamkha AJ, Aly AM, Mansour MA. Similarity solution for unsteady heat and mass transfer from a stretching surface embedded in a porous medium with suction/injection and chemical reaction effects. Chem Eng Commun 2010;197:846–58.

[61] Mukhopadhyay S. Unsteady boundary layer flow and heat transfer past a porous stretching sheet in presence of variable viscosity and thermal diffusivity. Int J Heat Mass Tran 2009;52:5213–7.

[62] Mukhopadhyay S. Effect of thermal radiation on unsteady mixed convection flow and heat transfer over a porous stretching surface in porous medium. Int J Heat Mass Tran 2009;52:3261–5.

[63] Mukhopadhyay S. Effects of slip on unsteady mixed convective flow and heat transfer past a porous stretching surface. Nucl Eng Des 2011;241:2660–5.

[64] Mukhopadhyay S, De PR, Bhattacharyya K, Layek GC. Casson fluid flow over an unsteady stretching surface. Ain Shams Eng J 2013;4:933–8.

[65] Bhattacharyya K, Mukhopadhyay S, Layek GC. Unsteady MHD boundary layer flow with diffusion and first order chemical reaction over a permeable stretching sheet with suction or blowing. Chem Eng Commun 2013;200:1–19.

[66] Pop I, Ingham DB. Convective Heat Transfer: Mathematical and Computational Modeling of Viscous Fluids and Porous Media. Oxford: Pergamon; 2001.

[67] Transport Phenomena in Porous Media. Ingham DB, Pop I, editors. vol II.

[68] Bejan A, Dincer I, Lorente S, Miguel AF, Reis AH. Porous and complex flow structures in modern technologies. New York: Springer; 2004.

[69] Ingham DB, Bejan A, Mamut E, Pop I, editors. Emerging technologies and techniques in porous media. Dordrecht: Kluwer; 2004.

[70] Ingham DB, Pop I, editors. Transport phenomena in porous media. Oxford: Elsevier; 2005, vol III.

[71] Vafai K. Handbook of porous media. Second ed New York: Taylor and Francis; 2005.

[72] Mukhopadhyay S, Layek GC. Radiation effect on forced convective flow and heat transfer over a porous plate in a porous medium. Meccanica 2009;44:587–97.

[73] Mukhopadhyay S, Layek GC. Effects of variable fluid viscosity on flow past a heated stretching sheet embedded in a porous medium in presence of heat source/sink. Meccanica 2012;44(4):863–76.

[74] Sajid M, Ali N, Javed T, Abbas Z. Stretching a curved surface in a viscous fluid. Chin Phys Lett 2010;024703.

[75] M. Sajid, N. Ali, T. Javed, and Z. Abbas, Flow of a micropolar fluid over a curved stretching sheet, J. Eng. Phys. Thermophys. (in press).

[76] Abbas Z, Naveed M, Sajid M. Heat transfer analysis for stretching flow over a curved surface with magnetic field. J Eng Thermophys 2013;22(4):337–45.

[77] Schlichting H. Boundary layer theory. McGraw-Hill; 1960.

[78] Hiemenz K. Die Grenzschict neinem in den gleichformigen flussigkeitsstrom eingetauchten geraden Kreiszylinder. Dinglers Polytech J 1911;326:321–410.

[79] Goldstein S. Modern development in fluid dynamics. London: Oxford University Press; 1938.

[80] Gersten K, Papenfuss HD, Gross JF. Influence of the Prandtl number on second-order heat transfer due to surface curvature at a three dimensional stagnation point. Int J Heat Mass Tran 1978;21:275–84.

[81] Liao SJ. A uniformly valid analytic solution of 2D viscous flow past a semi-infinite flat plate. J Fluid Mech 1999;385:101–28.

[82] Weidman PD, Putkaradze V. Axisymmetric stagnation flow obliquely impinging on a circular cylinder. Eur J Mech B/Fluids 2003;22:123–31.

[83] Stuart JT. The viscous flow near a stagnation point when the external flow has uniform vorticity. J Aerospace Sci 1959;26:124–5.

[84] Tamada KJ. Two-dimensional stagnation point flow impinging obliquely on a plane wall. J Phys Soc Jpn 1979;46:310–1.

[85] Takemitsu N, Matunobu Y. Unsteady stagnation-point flow impinging obliquely on an oscillating flat plate. J Phys Soc Jpn 1979;47:1347–53.

[86] Dorrepaal JM. An exact solution of the Navier-Stokes equation which describes non-orthogonal stagnation-point flow in two dimensions. J Fluid Mech 1986;163:141–7.

[87] Dorrepaal JM. Is two-dimensional oblique stagnation-point flow unique? Can Appl Math Q 2000;8:61–6.

[88] Labropulu F, Dorrepaal JM, Chandna OP. Oblique flow impinging on a wall with suction or blowing. Acta Mech 1996;115:15–25.

[89] Tilley BS, Weidman PD. Oblique two-fluid stagnation-point flow. Eur J Mech B/Fluids 1998;17:205–17.

[90] Kapur JN, Srivastava RC. Similar solutions of the boundary layer equations for power law fluids. Z Angew Math Phys 1963;14:383–8.

[91] Maiti MK. Axially-symmetric stagnation point flow of power law fluids. Z Angew Math Phys 1965;16:594–8.

[92] Koneru SR, Manohar R. Stagnation point flows of non-Newtonian power law fluids. Z Angew Math Phys 1968;19:84–8.

[93] Sapunkov YG. Similarity solutions of boundary layer of non-Newtonian fluids in magnetohydrodynamics. Mech Zidkosti I Gaza 1967;6:77–82 (in Russian).

[94] Djukic DS. Hiemenz magnetic flow of power-law fluids. Trans ASME J Appl Mech 1974;41:822–3.

[95] Mahapatra TR, Gupta AS. Heat transfer in stagnation-point flow towards a stretching sheet. Heat Mass Transfer 2002;38:517–21.

[96] Nazar R, Amin N, Filip D, Pop I. Stagnation point flow of a micropolar fluid towards a stretching sheet. Int J Non-Linear Mech 2004;39:1227–35.

[97] Lok YY, Amin N, Pop I. Comments on: steady two-dimensional oblique stagnation-point flow towards a stretching surface: M. Reza and A.S. Gupta. Fluid Dyn Res 2007;39:505–10.

[98] Reza M, Gupta AS. Steady two-dimensional oblique stagnation-point flow towards a stretching surface. Fluid Dyn Res 2005;37:334–40.

[99] Mahapatra TR, Dholey S, Gupta AS. Heat transfer in oblique stagnation-point flow of an incompressible viscous fluid towards a stretching surface. Heat Mass Transfer 2007;43:767–73.

[100] Paullet J, Weidman P. Analysis of stagnation point flow towards a stretching sheet. Int J Non-Lin Mech 2007;42:1084–91.

[101] Lok YY, Amin N, Pop I. Non-orthogonal stagnation point flow towards a stretching sheet. Int J Non-Linear Mech 2006;41:622–7.

[102] Lok YY, Amin N, Pop I. Mixed convection flow near a nonorthogonal stagnation point towards a stretching vertical plate. Int J Heat Mass Tran 2007;50:4855–63.

[103] Lok YY, Amin N, Campean D, Pop I. Steady mixed convection flow of a micropolar fluid near the stagnation point on a vertical surface. Int J Numer Method Heat Fluid Flow 2005;15:654–70.

[104] Lok YY, Amin N, Pop I. Mixed convection near a nonorthogonal stagnation point flow on a vertical plate with uniform surface heat flux. Acta Mech 2006;186:99–112.

[105] Ramachandran N, Chen TS, Armaly BF. Mixed convection in stagnation flows adjacent to a vertical Surfaces. ASME J Heat Transfer 1988;110:373–7.

[106] Ishak A, Jafar K, Nazar R, Pop I. MHD stagnation point flow towards a stretching sheet. Physica A 2009;388:3377–83.

[107] Mahapatra TR, Gupta AS. Magnetohydrodynamic stagnation-point flow towards a stretching sheet. Acta Mech 2001;152:191–6.

[108] Mahapatra TR, Nandy SK, Gupta AS. Magnetohydrodynamic stagnation-point flow of a power-law fluid towards a stretching surface. Int J Non-Lin Mech 2009;44:124–9.

[109] Ziabakhsh Z, Domairry G, Ghazizadeh HR. Analytical solution of the stagnation-point flow in a porous medium by using the homotopy analysis method. J Taiwan Inst Chem Eng 2009;40:91–7.

[110] Wu Q, Weinbaum S, Andreopoulos Y. Stagnation point flow in a porous medium. Chem Eng Sci 2005;60:123–34.

[111] Kechil SA, Hashim I. Approximate analytical solution for MHD stagnation-point flow in porous media. Comm Nonlinear Sci Numer Simulat 2009;14:1346–54.

[112] Vajravelu K, Van Gorder RA. Nonlinear flow phenomena and homotopy analysis: fluid flow and heat transfer. Beijing: Higher Education Press; 2012 and Springer-Verlag Berlin Heidelberg.

[113] Mukhopadhyay S, Vajravelu K. Diffusion of chemically reactive species in Casson fluid flow over an unsteady permeable stretching surface. J Hydrodyn 2013;25:591–8.

[114] Cortell R. Viscous flow and heat transfer over a non-linearly stretching sheet. Appl Math Comput 2007;184:864–73.

[115] Mahapatra TR, Nandy SK, Vajravelu K, Van Gorder RA. Stability analysis of the dual solutions for stagnation-point flow over a non-linearly stretching surface. Meccanica 2012;47:1623–32.

Flow past a shrinking sheet

3

Recently, the study of flow over a shrinking sheet has garnered considerable attention. The flow induced by a shrinking sheet is quite different from forward stretching flow (Miklavčič and Wang [1]). A stretching sheet would induce a far-field suction toward the sheet whereas in case of a shrinking sheet, the velocity moves toward a fixed point. The boundary layer flow over a shrinking surface is encountered in several technological processes. Such situations occur in polymer processing, manufacturing of glass sheets, paper production, in textile industries, and many others. Another example that belongs to this class of problems is the cooling of a large metallic plate in a bath, which may be an electrolyte. In this case, the fluid flow is induced as a result of shrinking of the plate. If the physical background of the flow is examined, then it can be observed that the vorticity generated due to the shrinking sheet is not confined within the boundary layer, and consequently, a situation appears where some other external force is needed that can help confine the vorticity inside the boundary layer. Only then is steady flow possible (Miklavčič and Wang [1]). Also, an external force may delay the boundary layer separation and maintain the structure of boundary layer. Thus steady flow exists only when a magnetic field is applied or an adequate suction on the boundary is imposed or if there are partial slips at the boundary.

Most of the available literature deals with the boundary layer flow past a linearly shrinking surface. However, shrinking of a sheet may not be linear in reality. A number of processes are thus available that use different shrinking velocities such as linear, power-law, and exponential. In Section 3.1, we shall outline the heat transfer characteristics for flow past a linearly shrinking sheet, whereas flow past a nonlinearly shrinking sheet is presented in Section 3.2. In Section 3.3, we discuss the mathematical formulation and the numerical solution of the nonlinear differential equations and relevant boundary conditions arising in the problem of flow past an exponentially shrinking surface. In all these sections, the flow and temperature fields are considered to be at steady state. However, because of a sudden shrinking of the flat sheet or by a step change of the temperature of the sheet, in some cases the flow field and heat and mass transfer can be unsteady. Accordingly, in Section 3.4, we report the situation when the shrinking force and surface temperature are varying with time. There are generally multiple ways to solve a given problem. With the help of similarity methods, the governing time-dependent boundary layer equations are transformed into a set of ordinary differential equations and finally the numerical solutions of the reduced equations are obtained. In all the above sections, we consider that the shrinking surface is flat. The shrinking surface may not be flat in some situations: The flow due to a curved shrinking surface may have applications in various technical items. Hence, in Section 3.5, a theoretical study is presented to obtain a numerical solution for the viscous flow over a curved shrinking surface. Stagnation-point flow has wider applications in different technological processes. Also, motivated by this, in Section 3.6, stagnation-point flow over a shrinking surface is analyzed in detail.

Fluid Flow, Heat and Mass Transfer at Bodies of Different Shapes. http://dx.doi.org/10.1016/B978-0-12-803733-1.00003-X

3.1 Flow past a linearly shrinking sheet

The boundary layer flow due to a shrinking sheet has attracted much attention because of its abnormal behavior. Miklavčič and Wang [1] pioneered the study of flow on a shrinking sheet. They found that the vorticity over the shrinking sheet is not confined within a boundary layer. To control the velocity of the shrinking sheet in the boundary layer, Miklavčič and Wang [1] imposed an adequate suction on the boundary whereas Wang [2] considered a stagnation flow. Since then, numerous studies emerged, investigating different aspects of this problem. Using a second-order slip flow model, Fang et al. [3] solved the viscous flow over a shrinking sheet analytically. They found that the solution has two branches, in a certain range of the parameters. Bhattacharyya et al. [4] observed that the velocity and thermal boundary layer thicknesses for the second solutions are always larger than those of the first solutions. Weidman et al. [5], Postelnicu and Pop [6], and Roşca and Pop [7] performed a stability analysis to show that the upper branch solutions are stable and physically realizable whereas the lower branch solutions are not stable and, therefore, not physically possible. Fang et al. [8] obtained an analytical solution for thermal boundary layers for a shrinking sheet. Hayat et al. [9] found an analytic solution for magnetohydrodynamic (MHD) flow of a second-grade fluid over a shrinking sheet. Later, using a homotopy analysis method (HAM), Hayat et al. [10] also obtained an analytical solution for the MHD rotating flow of a second-grade fluid past a porous shrinking sheet. Noor et al. [11] reported a series solution of MHD viscous flow due to a shrinking sheet using the Adomian decomposition method.

In this section, we present the results for flow past a linearly shrinking sheet in presence of temperature-dependent fluid viscosity and thermal diffusivity.

3.1.1 Mathematical analysis of the problem

Let (u, v) be the velocity components in the (x, y) directions, respectively. Then the continuity, momentum, and energy equations for steady two-dimensional flow of a viscous incompressible electrically conducting fluid past a heated linearly shrinking sheet in the presence of a uniform magnetic field of strength B_0 (imposed along the y-axis) can be written as

$$\frac{\partial u}{\partial x} + \frac{\partial v}{\partial y} = 0, \tag{3.1}$$

$$u\frac{\partial u}{\partial x} + v\frac{\partial u}{\partial y} = \frac{1}{\rho}\frac{\partial \mu}{\partial T}\frac{\partial T}{\partial y}\frac{\partial u}{\partial y} + \frac{\mu}{\rho}\frac{\partial^2 u}{\partial y^2} - \frac{\sigma B_0^2}{\rho}u, \tag{3.2}$$

$$u\frac{\partial T}{\partial x} + v\frac{\partial T}{\partial y} = \frac{\partial}{\partial y}\left(\kappa\frac{\partial T}{\partial y}\right). \tag{3.3}$$

Here T is the temperature, κ is the thermal diffusivity, and ρ is the fluid density (assumed constant), μ is the coefficient of fluid viscosity, and σ is the conductivity of the fluid. Because the variation of thermal conductivity (hence thermal diffusivity) and viscosity μ with temperature are quite significant, the thermal diffusivity κ and the fluid viscosity are assumed to vary with temperature. In the range of temperature considered (i.e. 0–23°C), the variation of both density ρ and specific heat (c_p) with temperature is negligible, and hence they are taken as constants.

We take the boundary conditions as

$$u = -cx, v = v_w, T = T_w \text{ at } y = 0, \tag{3.4a}$$

$$u \to \infty, T \to T_\infty \text{ as } y \to \infty, \tag{3.4b}$$

where $c > 0$ is the stretching rate of the sheet, v_w is the velocity at the wall, $v_w < 0$ corresponds to the case of suction and $v_w > 0$ for blowing, and T_w is the uniform wall temperature and T_∞ is the free-stream temperature.

We introduce the following relations

$$u = \frac{\partial \psi}{\partial y}, v = -\frac{\partial \psi}{\partial x}, \theta = \frac{T - T_\infty}{T_w - T_\infty} \tag{3.5}$$

where ψ is the stream function and θ is the dimensionless temperature.

Temperature-dependent fluid viscosity is given by (Batchelor [12]),

$$\mu = \mu^*[a + b(T_w - T)] \tag{3.6}$$

where μ^* is the constant value of coefficient of viscosity far away from the sheet and a, b are constants with $b > 0$.

The variation of thermal diffusivity with the dimensionless temperature is written as $\kappa = \kappa_0(1 + \lambda\theta)$ (Elbashbeshy [13]), where λ is a parameter that depends on the nature of the fluid and κ_0 is the value of thermal diffusivity at the temperature T_w.

This relation agrees well with that of Saikrishnan and Roy [14] and also with Bird et al. [15] (neglecting second- and higher-order terms).

With the help of the relations (3.5) and (3.6), equation (3.1) is automatically satisfied and the other equations (3.2)–(3.3) take the form

$$\frac{\partial \psi}{\partial y}\frac{\partial^2 \psi}{\partial x \partial y} - \frac{\partial \psi}{\partial x}\frac{\partial^2 \psi}{\partial y^2} = -A\nu^*\frac{\partial \theta}{\partial y}\frac{\partial^2 \psi}{\partial y^2} + \nu^*[a + A(1 - \theta)]\frac{\partial^3 \psi}{\partial y^3} - cM^2\frac{\partial \psi}{\partial y} \tag{3.7}$$

$$\& \frac{\partial \psi}{\partial y}\frac{\partial \theta}{\partial x} - \frac{\partial \psi}{\partial x}\frac{\partial \theta}{\partial y} = \kappa_0(1 + \lambda\theta)\frac{\partial^2 \theta}{\partial y^2} + \kappa_0\lambda\left(\frac{\partial \theta}{\partial y}\right)^2, \tag{3.8}$$

where $A = b(T_w - T_\infty)$, $\nu^* = \frac{\mu^*}{\rho}$, and $\frac{\sigma B_0^2}{\rho} = cM^2$, M being the Hartman number.

Boundary conditions become

$$\frac{\partial \psi}{\partial y} = -cx, \frac{\partial \psi}{\partial x} = -v_w, \theta = 1 \text{ at } y = 0, \tag{3.9a}$$

$$\& \frac{\partial \psi}{\partial y} \to 0, \theta \to 0 \quad \text{as} \quad y \to \infty. \tag{3.9b}$$

To find the similarity solutions of this relevant boundary value problem, the scaling group of transformations, a special form of Lie group transformations is employed.

Introducing the simplified form of Lie group transformations, namely, scaling group of transformations,

$$\Gamma : x^* = x e^{\varepsilon \alpha_1}, y^* = y e^{\varepsilon \alpha_2}, \psi^* = \psi e^{\varepsilon \alpha_3}, u^* = u e^{\varepsilon \alpha_4}, v^* = v e^{\varepsilon \alpha_5}, \theta^* = \theta e^{\varepsilon \alpha_6}. \tag{3.10}$$

Γ transforms the point $(x, y, \psi, u, v, \theta)$ to $(x^*, y^*, \psi^*, u^*, v^*, \theta^*)$.

Substituting (3.10) into equations (3.7)-(3.8) we get

$$e^{\varepsilon(\alpha_1 + 2\alpha_2 - 2\alpha_3)} \left(\frac{\partial \psi^*}{\partial y^*} \frac{\partial^2 \psi^*}{\partial x^* \partial y^*} - \frac{\partial \psi^*}{\partial x^*} \frac{\partial^2 \psi^*}{\partial y^{*2}} \right) = -A v^* e^{\varepsilon(3\alpha_2 - \alpha_3 - \alpha_6)} \frac{\partial \theta^*}{\partial y^*} \frac{\partial^2 \psi^*}{\partial y^{*2}}$$
$$+ v^*(a + A) e^{\varepsilon(3\alpha_2 - \alpha_3)} \frac{\partial^3 \psi^*}{\partial y^{*3}} - v^* A e^{\varepsilon(3\alpha_2 - \alpha_3 - \alpha_6)} \theta^* \frac{\partial^3 \psi^*}{\partial y^{*3}} - c M^2 e^{\varepsilon(\alpha_2 - \alpha_3)} \frac{\partial \psi^*}{\partial y^*}, \tag{3.11}$$

$$e^{\varepsilon(\alpha_1 + \alpha_2 - \alpha_3 - \alpha_6)} \left(\frac{\partial \psi^*}{\partial y^*} \frac{\partial \theta^*}{\partial x^*} - \frac{\partial \psi^*}{\partial x^*} \frac{\partial \theta^*}{\partial y^*} \right) = \kappa_0 (1 + \lambda \theta^* e^{-\varepsilon \alpha_6}) e^{\varepsilon(2\alpha_2 - \alpha_6)} \frac{\partial^2 \theta^*}{\partial y^{*2}}$$
$$+ \kappa_0 \lambda e^{2\varepsilon(\alpha_2 - \alpha_6)} \left(\frac{\partial \theta^*}{\partial y^*} \right)^2. \tag{3.12}$$

The system remains invariant under the transformation Γ if and only if the following relations are satisfied

$$\alpha_1 + 2\alpha_2 - 2\alpha_3 = 3\alpha_2 - \alpha_3 - \alpha_6 = 3\alpha_2 - \alpha_3 = 3\alpha_2 - \alpha_3 - \alpha_6 = \alpha_2 - \alpha_3 \tag{3.13a}$$

and

$$\alpha_1 + \alpha_2 - \alpha_3 - \alpha_6 = 2\alpha_2 - \alpha_6 = 2(\alpha_2 - \alpha_6). \tag{3.13b}$$

These relations give $\alpha_2 = 0 = \alpha_6$, $\alpha_1 = \alpha_3$.

The boundary conditions yield $\alpha_1 = \alpha_4$, $\alpha_5 = 0$.

Thus, Γ reduces to

$$\Gamma : x^* = x e^{\varepsilon \alpha_1}, y^* = y, \psi^* = \psi e^{\varepsilon \alpha_1}, u^* = u e^{\varepsilon \alpha_1}, v^* = v, \theta^* = \theta. \tag{3.14}$$

Expanding by Taylor's series, we have

$$x^* - x = x \varepsilon \alpha_1, y^* - y = 0, \psi^* - \psi = \psi \varepsilon \alpha_1,$$
$$u^* - u = u \varepsilon \alpha_1, v^* - v = 0, \theta^* - \theta = 0.$$

In terms of differentials, we have

$$\frac{dx}{\alpha_1 x} = \frac{dy}{0} = \frac{d\psi}{\alpha_1 \psi} = \frac{du}{\alpha_1 u} = \frac{dv}{0} = \frac{d\theta}{0}. \tag{3.15}$$

From the subsidiary equations (3.15), one can easily get on integration

$$\eta = y, \psi = xF(\eta), \theta = \theta(\eta), \tag{3.16}$$

where F is an arbitrary function of η. Equations (3.11)–(3.12) now become

$$F'^2 - FF'' = -A\upsilon^*\theta' F'' + \upsilon^*[a + A(1-\theta)]F''' - cM^2 F', \tag{3.17}$$

$$F\theta' + \kappa_0 \lambda \theta'^2 + \kappa_0 (1 + \lambda\theta)\theta'' = 0. \tag{3.18}$$

The boundary conditions now become

$$F'(\eta) = -c, F(\eta) = S, \theta(\eta) = 1 \text{ at } \eta = 0, \tag{3.19a}$$

$$F'(\eta) \to 0, \theta(\eta) \to 0 \text{ as } \eta \to \infty. \tag{3.19b}$$

Introducing $\eta = \upsilon^{*\alpha} c^\beta \eta^*, F = \upsilon^{*\alpha'} c^{\beta'} F^*, \theta = \upsilon^{*\alpha''} c^{\beta''} \theta^*$ in equations (3.17) and (3.18) we get

$$\alpha = \alpha' = \frac{1}{2}, \alpha'' = 0, \beta' = -\beta = \frac{1}{2}, \beta'' = 0.$$

Taking $\eta^* = \eta, F^* = f, \theta^* = \theta$, the above equations (3.17) and (3.18) and the boundary conditions (3.19a) and (3.19b) finally take the following form:

$$f'^2 - ff'' = -A\theta' f'' + [a + A(1-\theta)]f''' - M^2 f', \tag{3.20}$$

$$Pr f\theta' + \lambda\theta'^2 + (1 + \lambda\theta)\theta'' = 0, \tag{3.21}$$

$$f'(\eta) = -1, f(\eta) = S, \theta(\eta) = 1 \text{ at } \eta = 0, \tag{3.22a}$$

$$f'(\eta) \to 0, \theta(\eta) \to 0 \text{ as } \eta \to \infty. \tag{3.22b}$$

Here, $Pr = \dfrac{\upsilon^*}{\kappa_0}$ is the Prandtl number, and $S = \dfrac{\upsilon_w}{\sqrt{\upsilon^* c}}$. $S > 0 (\upsilon_w < 0)$ corresponds to suction whereas $S < 0$ (i.e., $\upsilon_w > 0$) corresponds to blowing.

3.1.2 Numerical results and discussion

We obtain numerical solutions to the above boundary value problem where equations (3.20)–(3.21) are solved as an initial value problem using the fourth-order Runge-Kutta method. The numerical results are obtained for a region $0 < \eta < \eta_\infty$, where $\eta_\infty = 7$. Dual solutions are obtained (numerically) for a specific region of the governing parameters. In all the numerical computations, we have taken $a = 1$.

Numerical results are plotted through graphs to visualize the effects of the pertaining parameters. In Fig. 3.1, the effects of viscosity variation parameter A on velocity and temperature profiles are exhibited. Fluid velocity decreases in the first branch (upper branch) solution with the increasing values of viscosity variation parameter A, whereas the opposite behavior is noted in the second branch (lower branch) solution

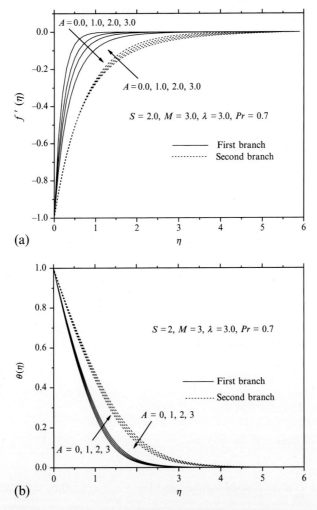

Figure 3.1 (a) Velocity and (b) temperature profiles for variable values of viscosity variation parameter A.

(Fig. 3.1(a)). The effect of A is more pronounced in the case of upper branch solution. Temperature increases with increasing values of A in the first branch solution, which is quite obvious as $A = b(T_w - T_\infty)$. But temperature decreases with increasing A in the second branch solution (Fig. 3.1(b)). In these investigations, high values of suction and magnetic parameter have been considered to confine the vorticity in the boundary layer region.

Figure 3.2(b) exhibits the nature of velocity profiles with the variation of thermal diffusivity parameter λ. With increasing λ, fluid velocity increases in the first branch solution but decreases in the second branch solution. Temperature also increases with the increasing values of λ in the case of the first branch, and the thermal boundary layer thickness increases with the increasing thermal diffusivity (Fig. 3.2(b)). However, the temperature decreases with the increasing values of thermal diffusivity parameter λ in the case of second branch solution (Fig. 3.2(b)).

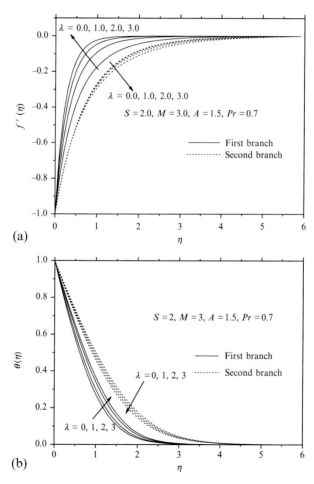

Figure 3.2 (a) Velocity and (b) temperature profiles for variable values of thermal diffusivity parameter λ.

3.2   Flow past a nonlinearly shrinking sheet

Flow due to a shrinking sheet is very important in the extrusion of sheet materials. Most of the available literature deals with the study of boundary layer flow over a shrinking surface where the velocity of the shrinking surface is assumed linearly proportional to the distance from the fixed origin. But realistically, a shrinking surface may not be linear.

Here we consider the velocity $u = -cx^n$ at $y = 0$, which is employed for positive odd integer values of n. It is clear that such a profile would fail for even integer values of n, as the flow at $y = 0$ would be in the wrong direction in case of $-\infty < x < 0$ (see Van Gorder and Vajravelu [16]). With the help of the modification provided in Van Gorder and Vajravelu [16], one can account for any values of $n \geq 1$, even for nonintegers. This allows us to consider a more general nonlinear power-law shrinking of the sheet. Here we consider the flow past a nonlinear shrinking sheet in presence of slip at the boundary. A more general shrinking velocity has been considered.

3.2.1   Formulation of the problem

Let (u, v) be the velocity components in the (x, y) directions, respectively. Then the continuity and momentum equations governing the flow of an incompressible viscous fluid past a nonlinearly shrinking sheet in the presence of slip at the boundary can be written as

$$\frac{\partial u}{\partial x} + \frac{\partial v}{\partial y} = 0, \tag{3.23}$$

$$u\frac{\partial u}{\partial x} + v\frac{\partial u}{\partial y} = \nu\frac{\partial^2 u}{\partial y^2}, \tag{3.24}$$

where $\nu = \dfrac{\mu}{\rho}$ is the kinematic viscosity, ρ is the fluid density, and μ is the coefficient of fluid viscosity.

The appropriate boundary conditions for the problem are given by

$$u = -c\,\mathrm{sgn}(x)|x|^n + N\nu\frac{\partial u}{\partial y}(-\infty < x < \infty), v = -V(x) \quad \text{at} \ \ y = 0, \tag{3.25a}$$

$$u \to 0 \quad \text{as} \ \ y \to \infty. \tag{3.25b}$$

Here c (>0) is a constant, n (>0) is a nonlinear shrinking parameter, $N = N_1|x|^{-\left(\frac{n-1}{2}\right)}$ is the velocity slip factor, which changes with x, and N_1 is the positive slip constant. The no-slip case is recovered for $N = 0$.

$V(x) > 0$ is the velocity of suction and $V(x) < 0$ is the velocity of blowing. $V(x) = V_0$ $\mathrm{sgn}(x)\,|x|^{\frac{n-1}{2}}$, a special type of velocity at the wall is considered, where V_0 is a constant.

It is to be noted that if $n > 1$, then N becomes singular at $x = 0$. As the boundary layer does not start at $x = 0$ but in the vicinity of $x = 0$, the solution for $n > 1$ is possible. We introduce the similarity variable and similarity transformations as

$$\eta = y\sqrt{\frac{c(n+1)}{2\nu}}|x|^{\frac{n-1}{2}}, u = c\,\mathrm{sgn}(x)|x|^n f'(\eta),$$

$$\nu = -\mathrm{sgn}(x)\sqrt{\frac{c(n+1)\nu}{2}}|x|^{\frac{n-1}{2}}\left\{f(\eta) + \left(\frac{n-1}{n+1}\right)\eta f'(\eta)\right\}, \tag{3.26}$$

and upon substitution of (3.26) in equations (3.24), (3.25a), and (3.25b), the governing equations and the boundary conditions reduce to

$$f''' + ff'' - \frac{2n}{n+1}f'^2 = 0, \tag{3.27}$$

$$f'(\eta) = -1 + Bf''(\eta), f(\eta) = S \quad \text{at} \quad \eta = 0 \tag{3.28a}$$

and

$$f'(\eta) \to 0 \quad \text{as} \quad \eta \to \infty. \tag{3.28b}$$

The prime denotes differentiation with respect to η, $S = \dfrac{V_0}{\sqrt{\dfrac{c\nu(n+1)}{2}}} > 0$ (or < 0) is the

suction (or blowing) parameter, and $B = N_1\sqrt{\frac{c\nu(n+1)}{2}}$ is the slip parameter.

3.2.2 Numerical solutions and discussion of the results

Numerical computations have been carried out using the fourth-order classical Runge-Kutta method with shooting technique. Numerical solution shows that dual solutions exist. We shall consider only the case of suction ($S > 0$) because injection will not permit the existence of solutions to the self-similar problem as it destroys the similarity flow over a shrinking sheet.

To assess the accuracy of the method, a comparison corresponding to the values of $[f''(0)]$ for a linear shrinking sheet in the case of the no-slip boundary condition is made with the results of Prasad et al. [17] and is presented in Table 3.1. From this table, it is clear that our results agree with their results up to the second decimal place.

In Fig. 3.3(a), velocity profiles are shown for different values of the nonlinear shrinking parameter n. Dual velocity profiles exist. Fluid velocity decreases with increasing values of n in the case of the first branch, whereas it increases finally for the second branch. One can notice that an increase in the value of the nonlinear shrinking parameter n helps to smoothen the velocity profile, which can be clearly

Table 3.1 Values of Skin-friction $f''(0)$ for different values of mass transfer parameter S in case of linear shrinking sheet in the absence of slip

	$S = 4.0$	$S = 4.5$	$S = 5.0$
Prasad et al. (2013) with $m = 1$	-1.038378	-2.228009	-3.798063
Present study with $n = 1$, $B = 0$	-1.037634	-2.226580	-3.796187

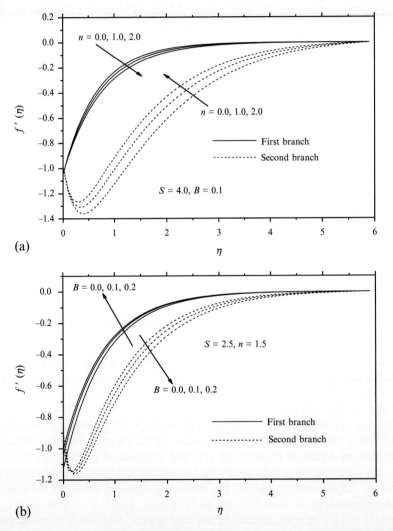

(a)

(b)

Figure 3.3 Effects of (a) nonlinear stretching parameter n and (b) slip parameter B on velocity profiles.

understood from the second branch solution (Fig. 3.3(a)). Fluid velocity increases with increasing slip parameter B in the first branch solution, but away from the sheet, the opposite behavior is noted in the case of the second branch solution (Fig. 3.3(b)). That is, the boundary layer thickness is less in the case of the first branch solution compared to that in the second branch solution. When slip occurs, the flow velocity near the sheet is no longer equal to the shrinking velocity of the sheet. With the increase in B, such slip velocity increases and consequently fluid velocity decreases because under the slip condition, the squeezing of the sheet can be only partly transmitted to the fluid. For increasing slip at the boundary (sheet), the generation of vorticity due to shrinking velocity is slightly reduced, and hence, with the imposed suction, that vorticity remains confined to the boundary layer region for larger shrinking velocity. Thus, one can say that because of increasing velocity slip, a steady solution is possible even for larger shrinking velocities.

3.3   Flow past an exponentially shrinking sheet

Flow past a shrinking sheet has gained considerable attention because of its interesting behaviors. The boundary layer flow over a shrinking surface is encountered in several technological processes. Fang and his coresearchers [18,19] discussed various aspects of flow past a shrinking sheet. Most of the earlier investigations deal with the study of boundary layer flow past a shrinking surface in which the velocity of the surface is assumed linearly proportional to the distance from the fixed origin. But the flow induced by an exponentially shrinking surface is a realistic one.

Flow and heat transfer characteristics past an exponentially shrinking sheet has not yet been addressed much. Bhattacharyya [20] investigated the boundary layer flow and heat transfer over an exponentially shrinking sheet. Wong et al. [21] studied the effects of viscous dissipation on boundary layer flow and heat transfer over an exponentially stretching/shrinking sheet in the presence of suction. Bhattacharyya and Pop [22] showed the effects of the external magnetic field on the flow over an exponentially shrinking sheet. Nadeem et al. [23] investigated MHD flow of Casson fluid induced by an exponentially shrinking sheet. They obtained analytical solutions by using ADM.

All of the previously mentioned studies assumed no-slip boundary conditions. In no-slip flow, as a requirement of continuum physics, the flow velocity is zero at a solid–fluid interface and the fluid temperature adjacent to the solid walls is equal to that of the solid walls. The nonadherence of the fluid to a solid boundary is known as velocity slip. It is a phenomenon that has been observed under certain circumstances (Yoshimura and Prudhomme [24]). The fluids exhibiting boundary slip find applications in technology such as in the polishing of artificial heart valves and internal cavities.

We shall present the results of flow and heat transfer past an exponentially shrinking permeable sheet in presence of a magnetic field, thermal radiation, and slip at the boundary.

3.3.1 Mathematical formulation

Consider the flow of an incompressible viscous electrically conducting fluid past a flat sheet coinciding with the plane $y=0$ in the presence of a magnetic field. The flow is confined to $y>0$. Two equal forces are applied along the x-axis toward the origin so that the surface shrinks keeping the origin fixed. Here the velocity of shrinking is of exponential order. A variable magnetic field $B(x)=B_0 e^{\frac{x}{2L}}$ is applied normal to the sheet, B_0 being a constant. These radiative effects have important applications in space technology and high-temperature processes. The continuity, momentum, and energy equations in the presence of thermal radiation are written as

$$\frac{\partial u}{\partial x} + \frac{\partial v}{\partial y} = 0, \tag{3.29}$$

$$u\frac{\partial u}{\partial x} + v\frac{\partial u}{\partial y} = \nu\frac{\partial^2 u}{\partial y^2} - \frac{\sigma B^2}{\rho}u, \tag{3.30}$$

$$u\frac{\partial T}{\partial x} + v\frac{\partial T}{\partial y} = \frac{\kappa}{\rho c_p}\frac{\partial^2 T}{\partial y^2} - \frac{1}{\rho c_p}\frac{\partial q_r}{\partial y} \tag{3.31}$$

where u and v are the components of velocity respectively in the x and y directions, $\nu = \frac{\mu}{\rho}$ is the kinematic viscosity, ρ is the fluid density, μ is the coefficient of fluid viscosity, q_r is the radiative heat flux, c_p is the specific heat at constant pressure, and κ is the thermal conductivity of the fluid.

Using Rosseland approximation for radiation (Brewster [25]), we can write

$$q_r = -\frac{4\sigma}{3k^*}\frac{\partial T^4}{\partial y} \tag{3.31a}$$

where σ is the Stefan-Boltzman constant, k^* is the absorption coefficient.

Assuming that the temperature difference within the flow is such that T^4 may be expanded in a Taylor series and expanding T^4 about T_∞, and neglecting higher orders, we get $T^4 \equiv 4T_\infty^3 T - 3T_\infty^4$. Therefore, equation (3) becomes

$$u\frac{\partial T}{\partial x} + v\frac{\partial T}{\partial y} = \frac{\kappa}{\rho c_p}\frac{\partial^2 T}{\partial y^2} + \frac{16\sigma T_\infty^3}{3\rho c_p k^*}\frac{\partial^2 T}{\partial y^2}. \tag{3.32}$$

The flow field can be changed significantly by the application of suction or injection (blowing) of a fluid through the bounding surface. In the design of thrust bearing and radial diffusers and thermal oil recovery, the process of suction and blowing is very important.

The appropriate boundary conditions for the problem are given by

$$u = U + N\nu\frac{\partial u}{\partial y}, v = -V(x), T = T_w + D\frac{\partial T}{\partial y} \text{ at } y=0, \tag{3.33a}$$

$$u \to 0, T \to 0 \quad \text{as} \quad y \to \infty. \tag{3.33b}$$

Here $U = -U_0 e^{\frac{x}{L}}$ is the shrinking velocity; $T_w = T_0 e^{\frac{x}{2L}}$ is the temperature at the sheet; U_0 and T_0 are the reference velocity and temperature, respectively; $N = N_1 e^{-\frac{x}{2L}}$ is the velocity slip factor, which changes with x; N_1 is the initial value of velocity slip factor; $D = D_1 e^{-\frac{x}{2L}}$ is the thermal slip factor, which also changes with x; and D_1 is the initial value of thermal slip factor. The no-slip case is recovered for $N = 0 = D$. $V(x) > 0$ is the velocity of suction, and $V(x) < 0$ is the velocity of blowing. $V(x) = V_0 e^{\frac{x}{2L}}$, a special type of velocity at the wall, is considered, where V_0 is the initial strength of suction.

Introducing the similarity variable and similarity transformations as

$$\eta = \sqrt{\frac{U_0}{2\nu L}} e^{\frac{x}{2L}} y, \, u = U_0 e^{\frac{x}{L}} f'(\eta),$$

$$v = -\sqrt{\frac{\nu U_0}{2L}} e^{\frac{x}{2L}} f(\eta) + \eta f'(\eta), \, T = T_0 e^{\frac{x}{2L}} \theta(\eta), \tag{3.34}$$

upon substitution of (3.34) in equations (3.30) and (3.32), the governing equations reduce to

$$f''' + f f'' - 2 f'^2 - M^2 f' = 0, \tag{3.35}$$

$$\left(1 + \frac{4}{3} R\right) \theta'' + Pr\left(f\theta' - f'\theta\right) = 0 \tag{3.36}$$

and the boundary conditions take the following form:

$$f'(\eta) = -1 + \lambda f''(\eta), \, f(\eta) = S, \theta(\eta) = 1 + \delta \theta'(\eta) \quad \text{at} \quad \eta = 0 \tag{3.37a}$$

and

$$f'(\eta) \to 0, \, \theta(\eta) \to 0 \quad \text{as} \quad \eta \to \infty. \tag{3.37b}$$

The prime denotes differentiation with respect to η, $M = \sqrt{\frac{2\sigma B_0^2 L}{\rho U_0}}$ is the magnetic parameter, $\lambda = N_1 \sqrt{\frac{U_0 \nu}{2L}}$ is the velocity slip parameter, $\delta = D_1 \sqrt{\frac{U_0}{2\nu L}}$ is the thermal slip parameter, $S = \dfrac{V_0}{\sqrt{\dfrac{U_0 \nu}{2L}}} > 0$ (or < 0) is the suction (or blowing) parameter, $R = \frac{4\sigma T_\infty^3}{\kappa k^*}$ is the radiation parameter, and $Pr = \frac{\mu c_p}{\kappa}$ is the Prandtl number.

3.3.2 Numerical solutions and analysis of the results

Using the shooting method, numerical calculations up to a desired level of accuracy are carried out for different values of dimensionless parameters of the problem under consideration for the purpose of illustrating the results graphically.

The shrinking rate in the exponential case is much faster than that of the linear case. Thus, the amount of vorticity generated due to exponential shrinking is greater than that of linear shrinking. So we have used magnetic field, suction, and slip to control the boundary layer flow. Fang and Zhang [26] showed that steady flow of Newtonian fluid past a shrinking sheet is possible only when the mass suction parameter is greater than or equal to 2. But Bhattacharyya [20] showed that if the mass suction parameter is greater than or equal to 2.266684, then only the steady flow due to an exponentially shrinking sheet is possible.

Figure 3.4(a) shows the velocity profiles for various values of the magnetic parameter M. A crossing-over point is noted in both cases of the first (upper) and second (lower) branch solutions. Behavior of fluid velocity is opposite before and after this crossing-over point. In the case of the first (upper) branch solution, fluid velocity initially decreases with increasing values of the magnetic parameter but it finally

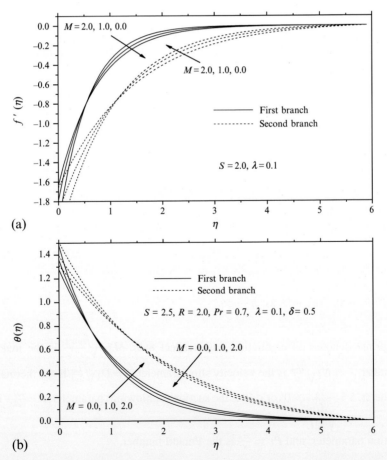

(a)

(b)

Figure 3.4 Effects of magnetic parameter M on (a) velocity and (b) temperature distributions.

increases with increasing M. But the opposite behavior is noted in the case of the second (lower) branch solution (Fig. 3.4(a)).

Though the temperature initially decreases with increasing M, it finally increases with increasing values of M for the first (upper) branch solution (see Fig. 3.4(b)). However, opposite trends for temperature profiles are noted in the case of the second (lower) branch solution.

In Fig. 3.5(a), velocity profiles are shown for different values of λ. The velocity curves show that the rate of transport decreases with the increasing distance (η) normal to the sheet. In all cases, the velocity vanishes at some large distance from the sheet (at $\eta = 6$).

With increasing λ, the streamwise component of the velocity is found to decrease for the first branch solution but the reverse trend is noted for the second branch solution. When slip occurs, the flow velocity near the sheet is no longer equal to the shrinking velocity of the sheet. With the increase in λ, such slip velocity increases, and consequently fluid velocity decreases. Further, with increasing λ, the temperature increases with λ for the upper branch solution (Fig. 3.5(b)), but a reverse effect of velocity slip is noted on the lower branch solution.

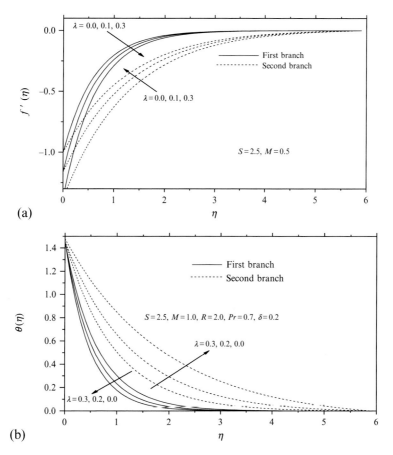

Figure 3.5 Effects of velocity slip parameter λ on (a) velocity and (b) temperature profiles.

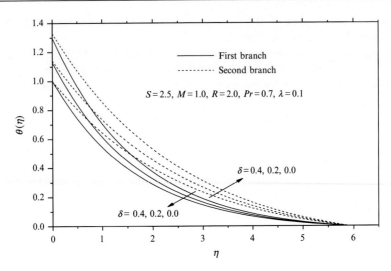

Figure 3.6 Effects of thermal slip parameter δ on temperature profiles.

Figure 3.6 depicts the effects of the thermal slip parameter δ on temperature. With increasing δ, temperature increases in the first branch solution but decreases in case of the second branch solution. With the increase of thermal slip parameter δ, heat is transferred to the fluid from the sheet, and so the temperature is found to increase.

3.4   Flow past an unsteady shrinking sheet

Most of the studies related to a shrinking surface are of the steady state. But in certain circumstances, the flow becomes time dependent and, consequently, it is necessary to consider the unsteadiness. The unsteady flow due to a shrinking sheet was studied by Fang et al. [27], and they numerically obtained the solution. The unsteady boundary layer flow on a shrinking sheet in an electrically conducting fluid under the effect of a transverse magnetic field of constant strength was considered by Merkin and Kumaran [28]. The unsteady three-dimensional viscous flow over a continuously permeable shrinking surface was analyzed by Bachok et al. [29]. Bhattacharyya [30] considered the effects of radiation and heat source/sink on an unsteady MHD boundary layer flow and heat transfer over a shrinking sheet with suction/injection. Fan et al. [31] investigated the unsteady stagnation-point flow and heat transfer toward a shrinking sheet using HAM, and Bhattacharyya [32] discussed the unsteady boundary layer stagnation-point flow over a shrinking sheet using a numerical method. The unsteady boundary layer flow of a nanofluid over a permeable stretching/shrinking sheet is studied by Bachok et al. [33]. We now present the flow past an unsteady shrinking sheet embedded in a porous medium.

3.4.1   Mathematical formulation of the problem

We consider laminar boundary-layer flow of viscous incompressible fluid over an unsteady permeable sheet continuously shrinking in its own plane with velocity $U(x,t)$ in a porous medium with permeability $k(t) = k_0(1 - \alpha t)$, where k_0 is the initial permeability. We assume that for time $t < 0$, the fluid flow is steady. The unsteady fluid flow starts at $t = 0$. The sheet emerges out of a slit at the origin $(x = 0, y = 0)$. In the case of polymer extrusion, the material properties of the sheet may vary with time.

The governing equations of the flow (using Darcy's law) are, in the usual notation,

$$\frac{\partial u}{\partial x} + \frac{\partial v}{\partial y} = 0, \tag{3.38}$$

$$\frac{\partial u}{\partial t} + u\frac{\partial u}{\partial x} + v\frac{\partial u}{\partial y} = v\frac{\partial^2 u}{\partial y^2} - \frac{v}{k}u. \tag{3.39}$$

Here u and n are the components of velocity, respectively, in the x and y directions, μ is the coefficient of fluid viscosity, ρ is the fluid density, and v is the kinematic viscosity of the fluid. Here, the linear Darcy term representing distributed body force due to porous media is retained whereas the nonlinear Forchheimer term is neglected.

The appropriate boundary conditions for the problem are given by

$$u = U(x, t) = \frac{-cx}{(1 - \alpha t)}, v = v_w(t) = \frac{-v_0}{\sqrt{(1 - \alpha t)}} \quad \text{at} \quad y = 0, \tag{3.40a}$$

$$u \to 0 \quad \text{as} \quad y \to \infty. \tag{3.40b}$$

The expressions for $U(x, t), k_1(t), v_w(t)$ are valid for time $t < \alpha^{-1}$. These forms are chosen in order to get a new similarity transformation, which transforms the governing partial differential equations into a set of ordinary differential equations, thereby facilitating the exploration of the effects of the controlling parameters. Here $v_0 (> 0)$ corresponds to suction whereas $v_0 (< 0)$ corresponds to blowing.

We now introduce the following relations for u, v as

$$u = \frac{\partial \psi}{\partial y}, v = -\frac{\partial \psi}{\partial x} \tag{3.41}$$

where ψ is the stream function.
We introduce

$$\eta = \sqrt{\frac{c}{v(1 - \alpha t)}}y, \psi = \sqrt{\frac{vc}{(1 - \alpha t)}}xf(\eta) \tag{3.42}$$

With the help of the above relations, the governing equations finally reduce to

$$A\left(\frac{\eta}{2}f'' + f'\right) + f'^2 - ff'' = f''' - k_1f', \tag{3.43}$$

where $A = \frac{\alpha}{c}$ is the unsteadiness parameter, $k_1 = \frac{v}{k_0c}$ is the permeability parameter, and $S = \frac{v_0}{\sqrt{vc}}$ is the suction/blowing parameter. $S > 0$ indicates suction, whereas $S < 0$ indicates blowing. Here we shall consider only the case of suction, as the flow over a shrinking sheet exists only for suction.

The boundary conditions then become

$$f' = -1, f = S \quad \text{at} \quad \eta = 0 \tag{3.44a}$$

and

$$f^/ \to 0 \text{ as } \eta \to \infty. \tag{3.44b}$$

3.4.2   Numerical solutions and discussion of the results

The self-similar equation (3.43) subject to the boundary conditions (3.44a-b) is solved numerically. To analyze the results, numerical data are presented graphically.

Figure 3.7 exhibits the velocity profiles for several values of unsteadiness parameter A. It is seen that the velocity along the sheet decreases with the increase of unsteadiness parameter A for the first branch solution, but velocity is found to increase with increasing A in case of the second branch solution (Fig. 3.7).

Figure 3.8 represents the effects of the permeability parameter on velocity profiles. With increasing permeability parameter k_1, fluid velocity increases in the case of the first branch solution but an opposite behavior is noted for the second branch (Fig. 3.8).

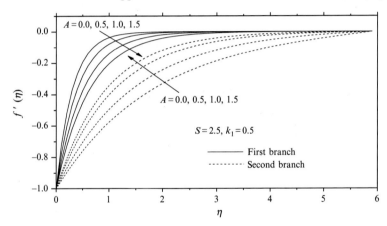

Figure 3.7 Effects of unsteadiness parameter A on velocity profiles.

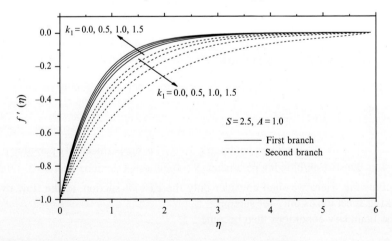

Figure 3.8 Effects of permeability parameter k_1 on velocity profiles.

3.5 Flow past a curved shrinking sheet

In all the above studies, the shrinking sheet is considered to be flat, and mathematical modeling is carried out using Cartesian coordinates. In the case of modeling of flow past a curved stretching sheet, one needs to consider curvilinear coordinates (Sajid et al. [34]). We present some important findings of such a problem in this section.

3.5.1 Formulation of the problem

Consider a two-dimensional boundary layer flow of an incompressible viscous fluid over a curved shrinking surface coiled in a cylinder of radius R about the curvilinear coordinates s and r.

Under boundary layer approximations, the governing equations for this type of flow are given by (Abbas et al. [35])

$$\frac{\partial}{\partial r}\{(R+r)v\} + R\frac{\partial u}{\partial s} = 0, \tag{3.45}$$

$$\frac{u^2}{r+R} = \frac{1}{\rho}\frac{\partial p}{\partial r}, \tag{3.46}$$

$$v\frac{\partial u}{\partial r} + \frac{Ru}{r+R}\frac{\partial u}{\partial s} + \frac{uv}{r+R} = -\frac{1}{\rho}\frac{R}{r+R}\frac{\partial p}{\partial s} + \nu\left(\frac{\partial^2 u}{\partial r^2} + \frac{1}{r+R}\frac{\partial u}{\partial r} - \frac{u}{(r+R)^2}\right). \tag{3.47}$$

Here, $\nu = \frac{\mu}{\rho}$ is the kinematic viscosity, μ is the coefficient of viscosity, ρ is the fluid density, r is the radial coordinate, s is the arc length coordinate, v and u are the velocity components in the r and s directions, respectively, and R is the distance of the sheet from the origin; large values of R correspond to small curvature.

The energy equation is given by

$$\rho c_p\left[r\frac{\partial T}{\partial r} + \frac{R}{r+R}u\frac{\partial T}{\partial s}\right] = \kappa\left[\frac{\partial^2 T}{\partial r^2} + \frac{1}{r+R}\frac{\partial T}{\partial r}\right], \tag{3.48}$$

where T is the temperature and κ is the thermal conductivity.

The appropriate boundary conditions are

$$u = as, v = -v_w, T = T_w \text{ at } r = 0, \tag{3.49a}$$

$$u \to 0, \frac{\partial u}{\partial r} \to 0, T \to T_\infty \text{ as } r \to \infty, \tag{3.49b}$$

where $a > 0$ is the stretching constant, $v_w = \dfrac{R}{r+R} v_0$ is the variable velocity at the wall and corresponds to suction when $v_0 > 0$, and v_0 being the initial value of the suction velocity.

Let us introduce the following similarity transformations:

$$u = -asf'(\eta), v = -\frac{R}{r+R}\sqrt{a\nu}f(\eta), p = \rho a^2 s^2 P(\eta),$$

$$\eta = \sqrt{\frac{a}{\nu}}r, \theta(\eta) = \frac{T-T_w}{T_w-T_\infty}.$$

(3.50)

Using these transformations, equation (3.45) is automatically satisfied, and equations (3.46)–(3.47) become

$$\frac{\partial P}{\partial \eta} = \frac{f'^2}{\eta+K},$$

(3.51)

$$\frac{2K}{\eta+K}P = f''' + \frac{1}{\eta+K}f'' - \frac{1}{(\eta+K)^2}f' - \frac{K}{\eta+K}f'^2 + \frac{K}{\eta+K}ff'' + \frac{K}{(\eta+K)^2}ff'.$$

(3.52)

Eliminating pressure P, we finally get the equation

$$f^{iv} + \frac{2}{\eta+K}f''' - \frac{1}{(\eta+K)^2}f'' + \frac{1}{(\eta+K)^3}f' - \frac{K}{(\eta+K)}\left(f'f'' - ff'''\right)$$

$$- \frac{K}{(\eta+K)^2}\left(f'^2 - ff''\right) - \frac{K}{(\eta+K)^3}ff' = 0,$$

(3.53)

and the energy equation takes the form

$$\theta'' + \frac{1}{(\eta+K)}\theta' + Pr\frac{K}{(\eta+K)}f\theta' = 0$$

(3.54)

$$f(0) = S, f'(0) = -1, \theta(0) = 1,$$

(3.55a)

$$f'(\infty) = 0, f''(\infty) = 0, \theta(\infty) = 0.$$

(3.55b)

Here $K = R\sqrt{\dfrac{a}{\nu}}$ is the curvature parameter, and $S = \dfrac{v_0}{\sqrt{a\nu}}(>0)$ is the suction parameter.

3.5.2 Numerical solutions and flow behaviors

Numerical results are obtained by solving the equations (3.53) and (3.54) subject to the boundary conditions (3.55a)–(3.55b) and are presented through graphs. Figure 3.9(a) shows that the horizontal component of fluid velocity increases with increasing values of curvature parameter K, whereas temperature decreases with increasing values of K (Fig. 3.9(b)).

Fluid velocity is found to increase with the increasing values of suction parameter S (Fig. 3.10(a)), whereas temperature is found to decreases with increasing S (Fig. 3.10(b)).

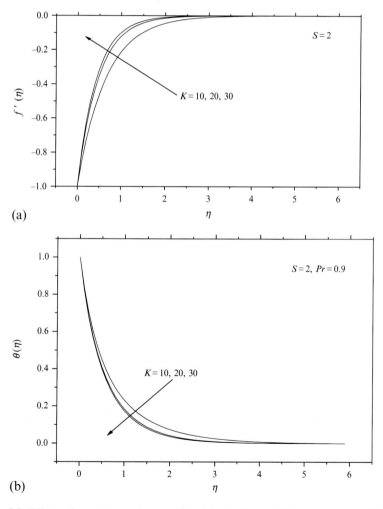

Figure 3.9 Effects of curvature parameter K on (a) velocity and (b) temperature profiles.

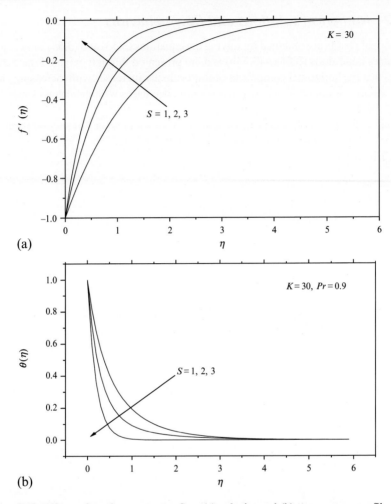

Figure 3.10 Effects of suction parameter S on (a) velocity and (b) temperature profiles.

3.6   Stagnation-point flow over a shrinking sheet

From the application points of view in different industries and technological processes, stagnation-point flow bears a great importance. Wang [36] first investigated the stagnation-point flow toward a shrinking sheet for both two-dimensional and axisymmetric cases. Ishak et al. [37] studied the steady boundary layer stagnation-point flow of a micropolar fluid over a shrinking sheet. Bhattacharyya and Layek [38] analyzed the effects of suction/blowing on the boundary layer stagnation-point flow and heat transfer toward a shrinking sheet in the presence of thermal radiation. Some other important characteristics of stagnation-point flow past a shrinking sheet can be found in the articles [39–45]. The MHD stagnation-point flow over a shrinking sheet was investigated by Mahapatra et al. [46] and Lok et al. [47]. Yacob et al. [48] investigated the melting heat transfer in steady boundary layer stagnation-point flow toward a horizontal linearly stretching/shrinking sheet in a micropolar fluid. The mass transfer characteristic in boundary layer stagnation-point flow toward a shrinking sheet was explained by Bhattacharyya [49] and Bachok et al. [50]. Fan et al. [51] investigated the unsteady stagnation-point flow and heat transfer toward a shrinking sheet using HAM. Bhattacharyya [52] discussed the unsteady boundary layer stagnation-point flow over a shrinking sheet using a numerical method. Rosali et al. [53] studied the stagnation-point flow and heat transfer past a shrinking sheet in a porous medium. The stagnation-point flow and heat transfer toward a shrinking sheet in a nanofluid were discussed by Nazar et al. [54]. The steady stagnation-point flow over an exponentially shrinking sheet was studied by Bhattacharyya and Vajravelu [55].

Motivated by the studies mentioned above, we shall present the numerical solutions for boundary layer flow and heat transfer of a viscous fluid past a shrinking sheet near a stagnation point. Actually, here we present the work of Bhattacharyya et al. [56]. The existence of dual solutions to the nonlinear equation is important from theoretical as well as practical point of view.

3.6.1 Mathematical formulation of the problem

Consider the stagnation-point flow of an incompressible viscous fluid past a linearly shrinking sheet. The equations of motion, in usual notation, for steady stagnation-point flow using the boundary layer approximations are

$$\frac{\partial u}{\partial x} + \frac{\partial v}{\partial y} = 0, \tag{3.56}$$

$$u\frac{\partial u}{\partial x} + v\frac{\partial u}{\partial y} = U\frac{dU}{dx} + \nu\frac{\partial^2 u}{\partial y^2}, \tag{3.57}$$

where u, v are the velocity components in the x and y directions, respectively, ν is the kinematic viscosity of the fluid, $U = U(x) = ax$ is the straining velocity of the stagnation-point flow, and a is a straining rate parameter.

The energy equation is given by

$$u\frac{\partial T}{\partial x} + v\frac{\partial T}{\partial y} = \frac{\kappa}{\rho c_p}\frac{\partial^2 T}{\partial y^2} \tag{3.58}$$

where ρ is the fluid density, T is the temperature, c_p is the specific heat, and κ is the thermal conductivity.

The boundary conditions are:

$$u = cx + L\left(\frac{\partial u}{\partial y}\right), v = 0, T = T_w \quad \text{at} \quad y = 0, \tag{3.59a}$$

$$u \to U(x) = ax, T \to T_\infty(x) \quad \text{as} \quad y \to \infty, \tag{3.59b}$$

where $c > 0$ is the stretching rate or $c < 0$ is the shrinking rate, T_w is the temperature of the sheet, and T_∞ is the free stream temperature with $T_w > T_\infty$.

We now introduce stream function ψ:

$$u = \frac{\partial \psi}{\partial y}, v = -\frac{\partial \psi}{\partial x}. \tag{3.60}$$

In view of (3.60), the above equation (3.56) is automatically satisfied and the other equations (3.57) and (3.58) become

$$\frac{\partial \psi}{\partial y}\frac{\partial^2 \psi}{\partial x \partial y} - \frac{\partial \psi}{\partial x}\frac{\partial^2 \psi}{\partial y^2} = U\frac{dU}{dx} + v\frac{\partial^3 \psi}{\partial y^3} \tag{3.61}$$

and

$$\frac{\partial \psi}{\partial y}\frac{\partial T}{\partial x} - \frac{\partial \psi}{\partial x}\frac{\partial T}{\partial y} = \frac{\kappa}{\rho c_p}\frac{\partial^2 T}{\partial y^2}. \tag{3.62}$$

The boundary conditions take the form

$$\frac{\partial \psi}{\partial y} = cx + L\frac{\partial^2 \psi}{\partial y^2}, \frac{\partial \psi}{\partial x} = 0, T = T_w \quad \text{at} \quad y = 0, \tag{3.63a}$$

$$\frac{\partial \psi}{\partial y} \to ax, T \to T_\infty \quad \text{as} \quad y \to \infty. \tag{3.63b}$$

Introducing the following dimensionless variables

$$\eta = y\sqrt{\frac{a}{v}}, \psi = \sqrt{avx}f(\eta), T = T_\infty + (T_w - T_\infty)\theta, \tag{3.64}$$

we obtain the equations

$$f''' + ff'' - f'^2 + 1 = 0, \tag{3.65}$$

$$\theta'' + Prf\theta' = 0. \tag{3.66}$$

The boundary conditions take the form

$$f(\eta) = 0, f'(\eta) = c/a + \delta f''(\eta), \theta(\eta) = 1 \quad \text{as} \quad \eta = 0, \tag{3.67a}$$

$$f'(\eta) \to 1, \theta(\eta) \to 0 \quad \text{as} \quad \eta \to \infty. \tag{3.67b}$$

Here c/a is the velocity ratio parameter and $\delta = L\sqrt{\dfrac{a}{\nu}}$ is the slip parameter.

The above self-similar equations subject to the boundary conditions are to be solved using the numerical method. The standard shooting technique with fourth-order Runge-Kutta method and Newton's method are used for finding approximate solutions of the above equations.

3.6.2   Numerical solutions and remarks

In order to validate the method used in this study and to judge the accuracy of the present analysis, comparisons with available results are made. The results in the absence of slip agree well with the results of Wang [36], which can be found from Table 3.2.

Hence, we believe that our results are accurate and reliable. Our results reveal that the existence and uniqueness of a solution depends on the slip parameter as well as the velocity ratio parameter.

It is observed that for no-slip boundary condition i.e. when $\delta = 0$, the solution is unique for $c/a > -1$, there exist dual solutions for $-1.2465 \le c/a \le -1$, and no similarity solution is found for $c/a < -1.2465$. These results are exactly the same as pointed out by Wang [36], who studied the stagnation-point flow toward a shrinking sheet with no-slip boundary conditions. Now for the slip parameter $\delta = 0.1$, the range of unique solution remains the same, that is, $c/a > -1$, but the dual solutions range significantly increases; that is, for this small increment of slip parameter, the range becomes $-1.31067 \le c/a \le -1$, and consequently for $c/a < -1.31067$ no

Table 3.2 Values of $f''(0)$ in the absence of slip at the boundary

Values of c/a	Present study		Wang [36]	
	First solution	Second solution	First solution	Second solution
−0.25	1.40224051		1.40224	
−0.50	1.49566948		1.49567	
−0.75	1.48929834		1.48930	
−1.0	1.32881689	0	1.32882	0
−1.15	1.08223164	0.11667340	1.08223	0.116702
1.2465	0.58429146	0.55428565	0.55430	

Reprinted from Ref. [4].

solution exists. Further enhancement in the value of the slip parameter δ causes the increment in the existence range of the similarity solution and the dual solutions range to increase. For $\delta = 0.5, 1, 1.5$, the ranges of c/a where dual solutions exist are $-1.69429 \leq c/a \leq -1$, $-2.33012 \leq c/a \leq -1$, $-3.02734 \leq c/a \leq -1$, respectively, and for all above values of δ, the solution is unique when $c/a > -1$. No similarity solution exists for $c/a < -1.69429$, $c/a < -2.33012$, $c/a < -3.02734$ when $\delta = 0.5$, 1, 1.5, respectively.

In Fig. 3.11(a), the dual velocity profiles confirm that the velocity decreases with increasing values of c/a in first solution and conversely for the second solution. Figure 3.11(b) gives an idea about the dual temperature profiles for various values

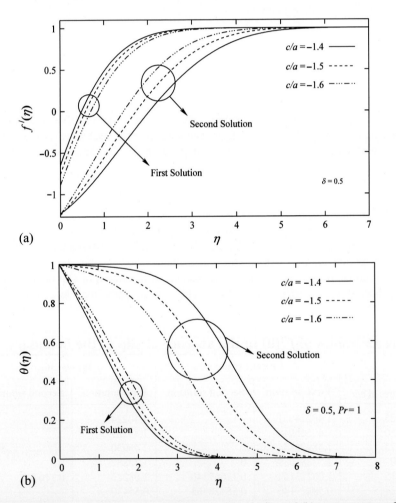

Figure 3.11 Effects of velocity ratio parameter on (a) velocity and (b) temperature profiles. Reprinted from Ref. [4].

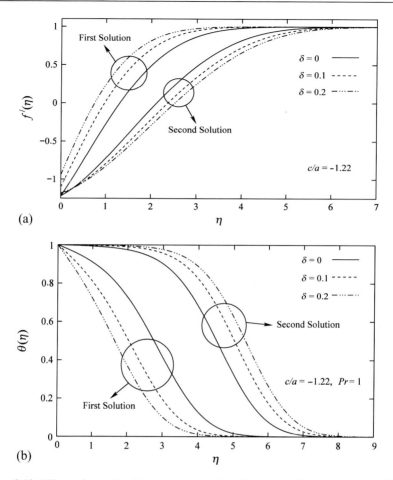

Figure 3.12 Effects of velocity slip parameter on (a) velocity and (b) temperature profiles. Reprinted from Ref. [4].

of c/a. The temperature at a point increases for increase in c/a in the first solution, whereas for the second solution it decreases.

With the increasing slip parameter δ, velocity increases in the first solution and decreases in the second solution except very near the sheet (Fig. 3.12(a)). With increasing slip value, the temperature at a fixed point decreases in the first solution case, but the reverse effect is observed in the second solution case (Fig. 3.12(b)).

References

[1] Miklavčič M, Wang CY. Viscous flow due to a shrinking sheet. Q Appl Math 2006;64: 283–90.
[2] Wang CY. Stagnation flow towards a shrinking sheet. Int J Non Lin Mech 2008;43: 377–82.

[3] Fang T, Yao S, Zhang J, Aziz A. Viscous flow over a shrinking sheet with a second order slip flow model. Comm Nonlinear Sci Numer Simulat 2010;15:1831–42.

[4] Bhattacharyya K, Mukhopadhyay S, Layek GC. Slip effects on boundary layer stagnation-point flow and heat transfer towards a shrinking sheet. Int J Heat Mass Tran 2011;54: 308–13.

[5] Weidman PD, Kubitschek DG, Davis AMJ. The effect of transpiration on self similar boundary layer flow over moving surfaces. Int J Eng Sci 2006;44:730–7.

[6] Postelnicu A, Pop I. Falkner–Skan boundary layer flow of a power-law fluid past a stretching wedge. Appl Math Comput 2011;217:4359–68.

[7] Roşca AV, Pop I. Flow and heat transfer over a vertical permeable stretching/shrinking sheet with a second order slip. Int J Heat Mass Tran 2013;60:355–64.

[8] Fang T, Liang W, Lee CF. A new solution branch for the Blasius equation—a shrinking sheet problem. Comp Math Appl 2008;56:3088–95.

[9] Hayat T, Abbas Z, Sajid M. On the analytic solution of magnetohydrodynamic flow of a second grade fluid over a shrinking sheet. J Appl Mech 2007;74:1165–71.

[10] Hayat T, Javed T, Sajid M. Analytic solution for MHD rotating flow of a second grade fluid over a shrinking surface. Phys Lett A 2008;372:3264–73.

[11] Noor NFM, Kechil SA, Hashim I. Simple non-perturbative solution for MHD viscous flow due to a shrinking sheet. Comm Nonlinear Sci Numer Simulat 2010;15:144–8.

[12] Batchelor GK. An introduction to fluid dynamics. London: Cambridge University Press; 1967, p. 597.

[13] Elbashbeshy EMA. Free convection flow with variable viscosity and thermal diffusivity along a vertical plate in the presence of the magnetic field. Int J Eng Sci 2000;38:207–13.

[14] Saikrishnan P, Roy S. Non-uniform slot injection (suction) into water boundary layers over (i) a cylinder and (ii) a sphere. Int J Eng Sci 2003;41:1351–65.

[15] Bird RB, Stewart WE, Lightfoot EN. Transport phenomena. New York: John Wiley and Sons; 1960.

[16] Van Gorder RA, Vajravelu K. A note on flow geometries and the similarity solutions of the boundary layer equations for a nonlinearly stretching sheet. Arch Appl Mech 2010;80: 1329–32.

[17] Prasad KV, Vajravelu K, Pop I. Flow and heat transfer at a nonlinearly shrinking porous sheet: the case of asymptotically large power-law shrinking rates. Int J Appl Mech Eng 2013;18:779–91.

[18] Fang T. Boundary layer flow over a shrinking sheet with power law velocity. Int J Heat Mass Tran 2008;51:5838–43.

[19] Fang T, Zhang J, Yao S. Slip magnetohydrodynamic viscous flow over a permeable shrinking sheet. Chin Phys Lett 2010;27:124702.

[20] Bhattacharyya K. Boundary layer flow and heat transfer over an exponentially shrinking sheet. Chin Phys Lett 2011;28(7):074701.

[21] Wong SW, Omar Awang MA, Ishak A, Pop I. Boundary layer flow and heat transfer over an exponentially stretching/shrinking permeable sheet with viscous dissipation. J Aerospace Eng 2014;27:26–32.

[22] Bhattacharyya K, Pop I. MHD boundary layer flow due to an exponentially shrinking sheet. Magnetohydrodynamics 2011;47:337–44.

[23] Nadeem S, Ul Haq R, Lee C. MHD flow of a Casson fluid over an exponentially shrinking sheet. Scientia Iranica B 2012;19:1550–3.

[24] Yoshimura A, Prudhomme RK. Wall slip corrections for Couette and parallel disc viscometers. J Rheologica 1998;32:53–67.

[25] Brewster MQ. Thermal radiative transfer properties. Hoboken, NJ: Wiley; 1972.

[26] Fang T, Zhang J. Closed-form exact solution of MHD viscous flow over a shrinking sheet. Comm Nonlinear Sci Numer Simulat 2009;14:2853–7.

[27] Fang T, Zhang J, Yao SS. Viscous flow over an unsteady shrinking sheet with mass transfer. Chin Phys Lett 2009;26(1):014703.

[28] Merkin JH, Kumaran V. The unsteady MHD boundary-layer flow on a shrinking sheet. Eur J Mech B Fluid 2010;29:357–63.

[29] Bachok N, Ishak A, Pop I. Unsteady three-dimensional boundary layer flow due to a permeable shrinking sheet. Appl Math Mech 2010;31:421–8.

[30] Bhattacharyya K. Effects of radiation and heat source/sink on unsteady MHD boundary layer flow and heat transfer over a shrinking sheet with suction/injection. Front Chem Sci Eng 2011;5:376–84.

[31] Fan T, Xu H, Pop I. Unsteady stagnation flow and heat transfer towards a shrinking sheet. Int Comm Heat Mass Tran 2010;37:1440–6.

[32] Bhattacharyya K. Dual solutions in unsteady stagnation-point flow over a shrinking sheet. Chin Phys Lett 2011;28:084702.

[33] Bachok N, Ishak A, Pop I. Unsteady boundary-layer flow and heat transfer of a nanofluid over a permeable stretching/shrinking sheet. Int J Heat Mass Tran 2012;55:2102–9.

[34] Sajid M, Ali N, Javed T, Abbas Z. Stretching a curved surface in a viscous fluid. Chin Phys Lett 2010;024703.

[35] Abbas Z, Naveed M, Sajid M. Heat transfer analysis for stretching flow over a curved surface with magnetic field. J Eng Thermophys 2013;22(4):337–45.

[36] Wang CY. Stagnation flow towards a shrinking sheet. Int J Nonlinear Mech 2008;43:377–82.

[37] Ishak A, Lok YY, Pop I. Stagnation-point flow over a shrinking sheet in a micropolar fluid. Chem Eng Comm 2010;197:1417–27.

[38] Bhattacharyya K, Layek GC. Effects of suction/blowing on steady boundary layer stagnation-point flow and heat transfer towards a shrinking sheet with thermal radiation. Int J Heat Mass Tran 2011;54:302–7.

[39] Nadeem S, Hussain A, Malik MY, Hayat T. Series solutions for the stagnation flow of a second-grade fluid over a shrinking sheet. Appl Math Mech 2009;30:1255–62.

[40] Kumari M, Nath G. Unsteady incompressible boundary layer flow of a micropolar fluid at a stagnation point. Int J Eng Sci 1984;22:755–68.

[41] Bachok N, Ishak A, Pop I. Melting heat transfer in boundary layer stagnation-point flow towards a stretching/shrinking sheet. Phy Lett A 2010;374:4075–9.

[42] Bachok N, Ishak A, Pop I. Stagnation-point flow over a stretchng/shrinking sheet in a nanofluid. Nanoscale Res Lett 2011;6:623.

[43] Bachok N, Ishak A, Pop I. Boundary layer stagnation-point flow toward a stretching/ shrinking sheet in a nanofluid. ASME J Heat Tran 2013;135:054501.

[44] Bachok N, Ishak A, Nazar R, Senu N. Stagnation-point flow over a permeable stretching/ shrinking sheet in a Cooper-Water nanofluid. Boundary Value Prob 2013;39:1–10.

[45] Bhattacharyya K, Arif MF, Ali Pk W. MHD boundary layer stagnation-point flow and mass transfer over a permeable shrinking sheet with suction/blowing and chemical reaction. Acta Tech 2012;57:1–15.

[46] Mahapatra TR, Nandy SK, Gupta AS. Momentum and heat transfer in MHD stagnation-point flow over a shrinking sheet. ASME J Appl Mech 2011;78:021015.

[47] Lok YY, Ishak A, Pop I. MHD stagnation-point flow towards a shrinking sheet. Int J Numer Mcth Hcat Fluid Flow 2011;21(1):61–72.

[48] Yacob NA, Ishak A, Pop I. Melting heat transfer in boundary layer stagnation-point flow towards a stretching/shrinking sheet in a micropolar fluid. Comput Fluids 2011;47:16–21.

[49] Bhattacharyya K. Dual solutions in boundary layer stagnation-point flow and mass transfer with chemical reaction past a stretching/shrinking sheet. Int Comm Heat Mass Tran 2011;38:917–22.

[50] Bachok N, Ishak A, Pop I. On the stagnation-point flow towards a stretching sheet with homogeneous-heterogeneous reactions effects. Comm Nonlinear Sci Numer Simulat 2011;16:4296–302.

[51] Fan T, Xu H, Pop I. Unsteady stagnation flow and heat transfer towards a shrinking sheet. Int Comm Heat Mass Tran 2010;37:1440–6.

[52] Bhattacharyya K. Dual solutions in unsteady stagnation-point flow over a shrinking sheet. Chin Phys Lett 2011;28:084702.

[53] Rosali H, Ishak A, Pop I. Stagnation point flow and heat transfer over a stretching/shrinking sheet in a porous medium. Int Comm Heat Mass Tran 2011;38:1029–32.

[54] Nazar R, Jaradat M, Arifin NM, Pop I. Stagnation-point flow past a shrinking sheet in a nanofluid. Central Eur J Phys 2011;9(5):1195–202.

[55] Bhattacharyya K, Vajravelu K. Stagnation-point flow and heat transfer over an exponentially shrinking sheet. Comm Nonlinear Sci Numer Simulat 2012;17:2728–34.

[56] Bhattacharyya K, Mukhopadhyay S, Layek GC. Slip effects on boundary layer stagnation-point flow and heat transfer towards a shrinking sheet. Int J Heat Mass Tran 2011;54(1-3):308–13.

Flow past a flat plate

4

Fluid flow over solid bodies occurs frequently, and it is responsible for numerous physical phenomena such as the drag force acting on automobiles, the lift developed by airplane wings, upward draft of rain, etc. So, developing a good understanding of external flow is very important in the design of several engineering systems. In an external flow, two distinct regimes exist:

(a) Flow outside the boundary layer: Here, the effect of viscosity is negligible. The velocities and pressures are affected by the physical presence of the body together with its boundary layer. Hence, the theories of ideal flow may be used.

(b) Flow immediately adjacent to the surface of the object: Here, viscosity effects are dominant. The notion of a boundary layer was first introduced by Prandtl [1] over a hundred years ago to explain the discrepancies between the theory of inviscid fluid flow and experiments.

Using scaling arguments, Prandtl obtained a set of equations for the conservation of momentum valid within the boundary layer. These equations are simpler than the full set of Navier-Stokes. Boundary-layer flows play an important role in several aspects of fluid mechanics as the whole dynamics is initiated from the boundary of the surface. The applicability of these boundary-layer equations was demonstrated by Blasius [2] for the flow past a flat plate at zero incidence, which showed excellent agreement with experimental data. Laminar flow over a flat plate is a fundamental problem of fluid mechanics. The study of the steady flow of viscous incompressible fluid over a flat plate has acquired significant attention because of its wide range of science and engineering applications.

4.1 Flow past a static horizontal plate

Laminar flow and heat transfer phenomena in porous media are of considerable interest due to their ever-increasing industrial applications and have important bearings on several technological processes. Transport processes through porous media play important roles in different applications, namely, in geothermal operations, petroleum industries, and many others. For this reason, considerable attention has been devoted to the study of boundary-layer flow behavior and heat transfer characteristics of a Newtonian fluid past a vertical plate embedded in a fluid-saturated porous medium (Pal and Shivakumara [3]). Processes involving heat and mass transfer in porous media are frequently encountered in the chemical industry, in reservoir engineering in connection with thermal recovery process, etc. (Murthy et al. [4]). A better understanding of convection through porous medium can benefit several areas like insulation design, grain storage, geothermal systems, heat exchangers, filtering devices, metal processing, catalytic reactors, etc. From both the theoretical and experimental standpoints, in the recent past, forced convection over a flat plate has been widely

Fluid Flow, Heat and Mass Transfer at Bodies of Different Shapes. http://dx.doi.org/10.1016/B978-0-12-803733-1.00004-1

studied. Soundalgekar et al. [5] investigated the combined free and forced convection flow past a semi-infinite plate with variable surface temperature. Watanabe and Pop [6] analyzed the Hall effects on magnetohydrodynamic (MHD) boundary-layer flow past a continuously moving plate. Elbashbeshy and Bazid [7] investigated the effects of non-Darcian porous medium on a moving plate. Anjali Davi and Kandasamy [8] reported the effects of chemical reaction on MHD flow past a plate. Recently, Damseh et al. [9] discussed the effects of magnetic fields and thermal radiation on forced convection flow. Mukhopadhyay and Layek [10] investigated the flow and heat transfer over a flat plate in a Darcy porous medium. Of late, Bhattacharyya and Layek [11] discussed the effects of suction/blowing on flow and mass transfer over a flat plate. Later on, Mukhopadhyay et al. [12] analyzed the effects of Darcy-Forchhimear porous medium on flow past a plate. We shall present here the results obtained by Mukhopadhyay and Layek [10].

We consider a forced convective, two-dimensional steady laminar boundary-layer flow of an incompressible, viscous liquid over a thin flat plate, embedded in porous medium (see Fig. 4.1).

In the analysis of flow in porous media, the differential equation governing the fluid motion is based on Darcy's law, which accounts for the drag exerted by the porous media. Thermal radiation is included in the energy equation. The governing equations of such a of flow are (with the application of Darcy's law)

$$\frac{\partial u}{\partial x} + \frac{\partial v}{\partial y} = 0, \tag{4.1}$$

$$u\frac{\partial u}{\partial x} + v\frac{\partial u}{\partial y} = v\frac{\partial^2 u}{\partial y^2} - \frac{v}{k}(u - u_\infty), \tag{4.2}$$

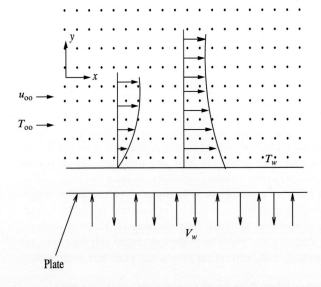

Figure 4.1 Sketch of the physical flow problem [10].

$$u\frac{\partial T}{\partial x} + v\frac{\partial T}{\partial y} = \frac{\kappa}{\rho c_p}\frac{\partial^2 T}{\partial y^2} - \frac{1}{\rho c_p}\frac{\partial q_r}{\partial y}. \tag{4.3}$$

Because the flow is laminar, the viscous dissipative heat is assumed to be negligible here. The x coordinate is measured from the leading edge of the plate, and the y coordinate is measured along the normal to the plate. Here u and v are the components of velocity, respectively, in the x and y directions, μ is the coefficient of fluid viscosity, ρ is the fluid density, $v = \mu/\rho$ is the kinematic viscosity, $k = k_0 x$ is the permeability of the porous medium, k_0 is the initial permeability, T is the temperature, κ is the thermal conductivity of the fluid, u_∞ is the free stream velocity, T_∞ is the free stream temperature, q_r is the radiative heat flux, and c_p is the specific heat at constant pressure.

Using Rosseland approximation for radiation (Brewster [13]), we can write

$$q_r = -\frac{4\sigma}{3k^*}\frac{\partial T^4}{\partial y},$$

where σ is the Stefan-Boltzmann constant and k^* is the absorption coefficient.

Assuming that the temperature difference within the flow is such that T^4 may be expanded in a Taylor series, expanding T^4 about T_∞ and neglecting higher orders, we get $T^4 \cong 4T_\infty^3 T - 3T_\infty^4$. Therefore, equation (4.3) becomes

$$u\frac{\partial T}{\partial x} + v\frac{\partial T}{\partial y} = \frac{\kappa}{\rho c_p}\frac{\partial^2 T}{\partial y^2} + \frac{16\sigma T_\infty^3}{3\rho c_p k^*}\frac{\partial^2 T}{\partial y^2}. \tag{4.4}$$

The appropriate boundary conditions for the problem are given by

$$u = 0, v = 0, T = T_w \text{ at } y = 0, \tag{4.5a}$$

$$u = u_\infty, T = T_\infty \text{ as } y \to \infty. \tag{4.5b}$$

Here T_w is the wall temperature, and T_∞ is the free stream temperature assumed constant with $T_w > T_\infty$.

For consideration of a porous plate, embedded in a porous medium, a suction/injection velocity through the plate is given. The boundary condition for the velocity component v then becomes $v = -v_w(x)$, where $v_w(x) = \frac{v_0}{\sqrt{x}}$ is the velocity of suction [$v_w(x) > 0$] or injection [$v_w(x) < 0$] of the fluid and v_0 is the constant value of suction/injection.

We now introduce the following relations for u, v:

$$u = \frac{\partial \psi}{\partial y}, v = -\frac{\partial \psi}{\partial x}, \tag{4.6}$$

where ψ is the stream function.

We introduce the following dimensionless variables:

$$\theta = \frac{T - T_\infty}{T_w - T_\infty},$$

(4.7)

and

$$\eta = y\sqrt{\frac{u_\infty}{\nu x}}, \psi = \sqrt{u_\infty \nu x} f(\eta).$$

(4.8)

Using the relations (4.6), (4.7), and (4.8) in the boundary-layer equation (4.2) and in the energy equation (4.4), we get the equations

$$f''' + \frac{1}{2}ff'' - \frac{1}{Da_x Re_x}\left(f' - 1\right) = 0,$$

(4.9)

$$\frac{1}{Pr}\left(1 + \frac{4}{3N}\right)\theta'' + \frac{1}{2}f\theta' = 0,$$

(4.10)

where $Da_x = \dfrac{k}{x^2} = \dfrac{k_0}{x}$ is the local Darcy number, $Re_x = \dfrac{u_\infty x}{\nu}$ is the local Reynolds number, $Pr = \dfrac{\mu c_p}{\kappa}$ is the Prandtl number, and $N = \dfrac{\kappa k^*}{4\sigma T_\infty^3}$ is the radiation parameter.

Equation (4.9) can be written as

$$f''' + \frac{1}{2}ff'' - k_1\left(f' - 1\right) = 0,$$

(4.11)

where $k_1 = \dfrac{1}{Da_x Re_x} = \dfrac{\nu}{k_0 u_\infty}$ is the parameter of the porous medium (Damesh [14]).

In view of these, the boundary conditions finally become

$$f' = 0, f = S, \theta = 1 \text{ at } \eta = 0$$

(4.12a)

and

$$f' = 1, \theta = 0 \text{ at } \eta \to \infty,$$

(4.12b)

where $S = \dfrac{2v_0}{\sqrt{u_\infty \nu}}$ is the suction ($v_0 > 0$) / injection ($v_0 < 0$) parameter.

When $k_1 = 0$, the equation (4.11) reduces to the equation of boundary-layer flow on a flat plate at zero incidence (Batchelor [15]).

4.1.1 Numerical solutions and discussion of the results

To get a clear insight of the physical problem, numerical computations have been carried out for various values of the parameters involved in the problem, and the numerical results are presented in Fig. 4.2.

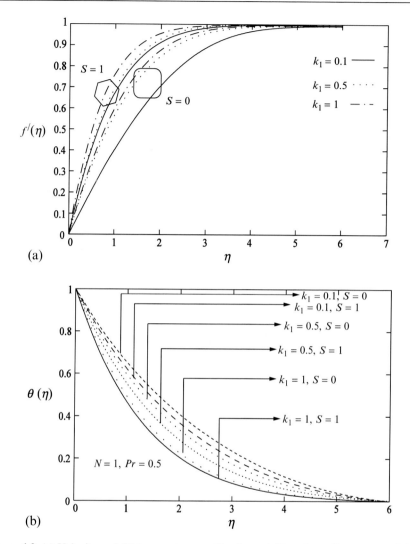

Figure 4.2 (a) Velocity and (b) temperature profiles for variable values of parameter of the porous medium [10].

Figure 4.2(a) illustrates the effects of the permeability of the medium on a velocity field in the absence ($S=0$) and presence ($S=1$) of suction. This figure shows the variations of velocity profiles for increasing values of the porous medium parameter k_1 when the other parameters N and Pr are kept constant (i.e., $N=1$ and $Pr=0.5$). Clearly, as k_1 increases, the peak value of velocity tends to increase. It clearly indicates that the thickness of the velocity boundary layer decreases. In this case, horizontal velocity is found to increase with the increasing values of the porous medium parameter. With a rise in permeability of the medium, the regime becomes more porous. As a consequence, the Darcian body force decreases in magnitude (as it is inversely proportional to the permeability). The Darcian resistance acts to decelerate the fluid particles in continua. This resistance diminishes as the permeability of the medium increases. So, progressively less drag is experienced by the flow, and flow retardation is thereby decreased. Hence, the velocity of the fluid increases as the parameter k_1 increases. On the other hand, by sucking fluid particles through a porous wall, the growth of the fluid boundary layer is reduced. So in the presence of suction, fluid velocity is found to be much more suppressed. Because the effect of suction is to suck away the fluid near the wall, the momentum boundary layer is reduced as a result of suction ($S > 0$). Consequently, the velocity increases. Hence, the velocity gradient and the skin friction increase with increasing S ($S > 0$). The effect of injection ($S < 0$) is opposite to that of suction ($S > 0$). The temperature profiles for several increasing values of k_1 taking $N = 1$ and $Pr = 0.5$ in the absence of suction ($S = 0$) and in the presence of suction ($S = 1$) are exhibited in Fig. 4.2(b). From this figure, it is found that the temperature decreases with the increase of k_1.

The rate of heat transfer (the thermal boundary-layer thickness becomes thinner) is enhanced when the velocity boundary-layer thickness decreases. Temperature is also found to decrease with the increase of η until it vanishes at $\eta = 6$. In the presence of suction, the temperature profiles are found to be suppressed as the growth of thermal boundary layer is reduced by sucking the fluid particles through the porous plate. The thermal boundary-layer thickness decreases with the suction parameter S, which causes an increase in the rate of heat transfer. The explanation for such behavior is that the fluid is brought closer to the surface and reduces the thermal boundary-layer thickness.

4.2   Flow past a moving horizontal plate

The problem of viscous boundary-layer flow on a moving flat plate is a classical problem, and it has been considered by many researchers. The phenomenon of boundary-layer behavior over a moving plate in a parallel free stream has several practical applications such as the aerodynamic extrusion of plastic sheets, the cooling of an infinite metallic plate in a cooling bath, the boundary layer along material handling conveyers, the boundary layer along a liquid film in condensation processes, paper production, etc. (Mukhopadhyay et al. [16]). In the case of boundary-layer flow due to a moving plate, the distribution of solute undergoing chemical reaction has many practical applications in the metallurgy industry, filaments drawn through a electrically conducting fluid, and also for the purification of molten metals (Mukhopadhyay et al. [16]). The momentum, heat, and mass transfer problems with moving-wall and with stream were studied by several researchers. Recently, by extending the work of Cortell [17], Ishak et al. [18] showed that dual solutions exist when the surface moves in an opposite direction to that of the free stream. Mukhopadhyay et al. [18] extended the work of Cortell [17] and Ishak et al. [18] by considering the concentration equation with a first-order reaction. To obtain the self-similar solutions, a space-dependent reaction rate is considered. A reaction is said to be first order if the rate of reaction is directly proportional to concentration itself. In this section, we report the results obtained by Mukhopadhyay et al. [16]. They considered the uniform wall concentration at the boundary.

Consider a two-dimensional steady laminar boundary-layer flow of an incompressible, viscous fluid with first-order chemical reaction between a flat moving surface with constant velocity U_w and a free stream moving in the same or opposite direction with velocity U_∞. The x-axis extends parallel to the surface whereas the y-axis extends upwards, normal to the surface.

The governing equations for boundary-layer flow and mass transfer are

$$\frac{\partial u}{\partial x} + \frac{\partial v}{\partial y} = 0, \tag{4.13}$$

$$u\frac{\partial u}{\partial x} + v\frac{\partial u}{\partial y} = \nu\frac{\partial^2 u}{\partial y^2}, \tag{4.14}$$

$$u\frac{\partial C}{\partial x} + v\frac{\partial C}{\partial y} = D\frac{\partial^2 C}{\partial y^2} - k(C - C_\infty), \tag{4.15}$$

where u and v are the components of velocity respectively in the x and y directions, $\nu = \mu/\rho$ is the kinematic viscosity, μ is the viscosity of the fluid, ρ is the fluid density, C is the species concentration in the fluid, D is the molecular diffusivity of chemically reactive species, and k ($k = \frac{Lk_0}{2x}$, k_0 is a constant having the same dimension as k and L is the characteristic length) is the space-dependent rate of chemical conversion of the first-order homogeneous and irreversible reaction. The diffusing species can either be destroyed or generated in the homogeneous reaction. The concentration of the reactant is maintained at a prescribed constant value C_w at the surface and is assumed to vanish far away for the surface.

The appropriate boundary conditions for the problem are

$$u = U_w, v = 0, C = C_w \text{ at } y = 0, \tag{4.16a}$$

$$u \rightarrow U_\infty, C \rightarrow C_\infty \text{ as } y \rightarrow \infty. \tag{4.16b}$$

Here C_w is the uniform surface concentration and C_∞ is the concentration far away from the surface. We assume $C_w > C_\infty$.

4.2.1 Solutions and discussion of the results

With the help of a composite velocity $U = U_w + U_\infty$ (U_w is the velocity of the surface and U_∞ is the free stream velocity), we now introduce the following dimensionless variables

$$\eta = y\sqrt{\frac{U}{2\nu x}}, u = Uf'(\eta), v = U\frac{\eta f'(\eta) - f(\eta)}{\sqrt{2Re_x}}, \tag{4.17}$$

$$\phi = \frac{C - C_\infty}{C_w - C_\infty}, \tag{4.18}$$

where $Re_x = \frac{Ux}{\nu}$ is the local Reynolds number.

Using the relations (4.17) and (4.18) in the boundary layer equation (4.14) and in the concentration equation (4.15), we get the following equations:

$$f''' + ff'' = 0, \tag{4.19}$$

$$\frac{\phi''}{Sc} + f\phi' - \beta\phi = 0, \tag{4.20}$$

where $Sc = \frac{\nu}{D}$ is the Schmidt number, $\beta = \frac{Lk_0}{U}$ is the reaction rate parameter, $\beta > 0$ indicates destructive chemical reaction, $\beta < 0$ denotes generative chemical reaction, and $\beta = 0$ represents the nonreactive species. The boundary conditions (4.16a) and (4.16b) in dimensionless form are

$$f' = 1 - r, f = 0, \phi = 1 \quad \text{at} \quad \eta = 0, \tag{4.21a}$$

$$f' = r, \phi = 0 \quad \text{as} \quad \eta \to \infty, \tag{4.21b}$$

where $r = \frac{U_\infty}{U}$ is the velocity ratio parameter.

Equations (4.19) and (4.20) along with boundary conditions are solved numerically by the shooting method (for details, see Mukhopadhyay et al. [19]).

As is often the case for nonlinear problems in fluid mechanics, exact solutions to the governing equations (4.19)–(4.21) are not always possible. However, because the equation is related to the classical Blasius problem (when $r = 1$, (4.19) and (4.21) become the Blasius problem; see, e.g., [20,21]), there is hope to find certain analytical approximations to the solutions. Such solutions are useful, as they often provide greater insight into the solutions than do purely numerical results.

Solutions to the classical Blasius problem have been treated frequently in the literature (see [20,21] and the references therein). Let us write $r = 1 + \varepsilon$: When $\varepsilon = 0$ we recover the classical Blasius problem. Given such a classical solution, we may, in principle, calculate a solution for (4.19) given (4.21) for small ε. Assume a solution f to (4.19) takes the form $f(\eta) = \hat{f}(\eta) + \varepsilon g(\eta)$, where $\hat{f}$ is a given solution to the classical Blasius problem (i.e., the case where $\varepsilon = 0$) and g takes into account the deviation from $\varepsilon = 0$. Here we have neglected higher-order terms in ε. We find that g is governed by the linear boundary value problem

$$g''' + \hat{f}g'' + \hat{f}''g = 0, \tag{4.22}$$

$$g = 0, g' = -1 \quad \text{at} \quad \eta = 0, \tag{4.23}$$

$$g' = 1 \quad \text{as} \quad \eta \to \infty. \tag{4.24}$$

Integrating (4.22) three times, and making use of the relevant boundary conditions (4.23) and (4.24), we obtain the integral equation

$$g(\eta) = \frac{g''(0)}{2}\eta^2 - \eta + \int_0^\eta K(\eta, \xi)g(\xi)d\xi, \tag{4.25}$$

where the kernel $K(\eta, \xi)$ is defined in terms of $\hat{f}$ and is given by

$$K(\eta, \xi) = \hat{f}(\xi) + 2(\eta - \xi)\hat{f}'(\xi) + (\eta - \xi)^2 \hat{f}''(\xi). \tag{4.26}$$

Furthermore, the value $g''(0)$ may be calculated by

$$g''(0) = \lim_{\eta \to \infty} \left\{ (\hat{f}(\eta)g(\eta))' - 2\int_0^\eta \hat{f}'(\xi)g'(\xi)d\xi \right\} = L. \tag{4.27}$$

From the theory of linear integral equations, (4.25) yields

$$g(\eta) = \frac{L}{2}\eta^2 - \eta + \int_0^\eta \left\{ \frac{L}{2}\xi^2 - \xi \right\} K(\eta, \xi)\exp\left(\int_\xi^\eta K(\chi, \chi)d\chi\right)d\xi, \tag{4.28}$$

where the right-hand side is given explicitly in terms of the known function $\hat{f}$. Thus, for small ε, we have (see Mukhopadhyay et al. [16])

$$f(\eta) = \hat{f}(\eta) +$$
$$\varepsilon\left\{ \frac{L}{2}\eta^2 - \eta + \int_0^\eta \left\{ \frac{L}{2}\xi^2 - \xi \right\} K(\eta, \xi)\exp\left(\int_\xi^\eta K(\chi, \chi)d\chi\right)d\xi \right\}. \tag{4.29}$$

Once a solution f to (4.19) and (4.21) is obtained, it may be plugged into (4.20) so that a solution for the nondimensional chemical concentration ϕ can be obtained. Although (4.20) is linear in ϕ, it has variable coefficients given in terms of f and, therefore, the solution procedure is nontrivial. Let us make the substitution

$$\phi(\eta) = \exp\left(-\int_0^\eta P(\xi)d\xi\right). \tag{4.30}$$

Then, equation (4.20) becomes the Riccati equation

$$\frac{1}{Sc}P' = \frac{1}{Sc}P^2 - f(\eta)P - \beta. \tag{4.31}$$

Here we assume that a solution f is given and plugged into (4.20). Note that, in the large Sc limit (i.e., $Sc \to \infty$), we get $P = -\beta/f(\eta)$. This yields the exact solution

$$\phi(\eta) = \exp\left(\beta\int_0^\eta \frac{1}{f(\xi)}d\xi\right). \tag{4.32}$$

Note that this is completely consistent with the boundary conditions (4.21a) and (4.21b). In the small Sc limit (i.e., $Sc \to 0$), we have from (4.20) that $\phi'' = 0$ which implies a linear solution. Indeed, we see from (4.31) that if we assume a small-Sc expansion, we obtain

$$\phi(\eta) = \exp\left(\int_0^\eta \left\{\frac{1}{C+\xi} + O(Sc)\right\} d\xi\right) = \frac{C+\eta}{C} \exp\left(-\int_0^\eta O(Sc) d\xi\right), \qquad (4.33)$$

where C is a constant of integration. There is no constant C such that both $\phi(0) = 1$ and $\phi \to 0$ as $\eta \to \infty$. Hence, there exists no linear solution satisfying both boundary conditions given in (4.21a) and (4.21b), so the small Sc limit is a singular limit that does not agree qualitatively with physical solutions. This is not to say that solutions do not exist for small Sc. Rather, since perturbation for small Sc involves computing a solution for the $Sc = 0$ case, which is itself inconsistent with the boundary conditions, the $Sc \to 0$ limit is a singular limit and regular perturbation cannot be used there.

Let us consider a perturbative solution in the large Sc regime. Assume a solution to (4.31) takes the form

$$P(\eta) = P_0(\eta)\left\{1 + \frac{1}{Sc}P_1(\eta) + \frac{1}{Sc^2}P_2(\eta) + O\left(\frac{1}{Sc^3}\right)\right\}. \qquad (4.34)$$

We find that

$$P_0 = -\frac{\beta}{f}, \ P_1 = \frac{f' - \beta}{f^2}, \ P_2 = \frac{(f' - \beta)(3f' - \beta) - ff''}{f^4}. \qquad (4.35)$$

From here, we find that

$$\phi(\eta) = \exp\left(\int_0^\eta \frac{\beta}{f(\xi)} \left\{ \begin{array}{l} 1 + \dfrac{1}{Sc^2}\dfrac{(f'(\xi) - \beta)(3f'(\xi) - \beta) - f(\xi)f''(\xi)}{f(\xi)^4} \\[2mm] + \dfrac{1}{Sc}\dfrac{f'(\xi) - \beta}{f(\xi)^2} \end{array} \right\} d\xi \right) \qquad (4.36)$$

up to third order in Sc^{-1}. This solution has the desired asymptotic behavior when $\beta < 0$. Further, for $\beta > 0$, similar results can be obtained upon taking the positive exponential power in (4.30). In the special case of $\beta = 0$, we can solve (4.20) exactly, and using (4.21a) we obtain

$$\phi(\eta) = 1 + \phi'(0)\int_0^\eta \exp\left(-\int_0^\xi f(\chi) d\chi\right) d\xi. \qquad (4.37)$$

Making use of (4.21b), we have

$$\phi'(0) = -\left[\int_0^\infty \exp\left(-\int_0^\xi f(\chi)d\chi\right)d\xi\right]^{-1}. \tag{4.38}$$

Then,

$$\phi(\eta) = 1 - \frac{\int_0^\eta \exp\left(-\int_0^\xi f(\chi)d\chi\right)d\xi}{\int_0^\infty \exp\left(-\int_0^\xi f(\chi)d\chi\right)d\xi}. \tag{4.39}$$

In order to get a clear insight of the physical problem, numerical computations have been carried out using a shooting method for various values of the parameters involved in this problem.

The effects of the velocity ratio parameter r on the velocity is exhibited in Fig. 4.3(a). From this figure, the dual velocity profiles can be found, and it is observed that with increasing velocity ratio parameter, the momentum boundary layer thickness increases in the upper branch solution, and for the lower branch solution it decreases with increasing r.

Next we consider the effects of the velocity ratio parameter on the concentration field in the presence of destructive and constructive chemical reaction separately. For the upper branch solution, the effect of increasing r is to increase the concentration (Fig. 4.3(b)) in the boundary layer region when a destructive chemical reaction takes place. But in this case, the opposite effect is noted on the lower branch solution (Fig. 4.3(b)). Consequently, the solute boundary-layer thickness is found to increase with increasing r for the upper branch, and in the case of the lower branch, it decreases. In the presence of a destructive chemical reaction, the concentration gradient also increases for the upper branch solution, but the opposite behavior is noted for the lower branch solution. When the constructive chemical reaction takes place, then the concentration increases with increasing r for the upper branch solution, but the concentration is found to decrease with increasing r for the lower branch solution (Fig. 4.3(c)). In this case, the concentration gradient at the wall for the upper branch solution is negative but it takes positive values for the lower branch solution; that is, there is a mass transfer from the fluid to the plate. That is why the overshoot in the concentration profiles (lower branch) is noted near the surface.

The numerical solutions along with the perturbation solutions are plotted in Fig. 4.4(a) and (b). In Fig. 4.4(a), we see that the perturbation solutions agree well with the numerical solutions at $r = 1.2$ (corresponding to $\varepsilon = 0.2$), whereas the solutions begin to disagree for larger values of r. When $r = 1.4$, the perturbation solutions in presence of constructive chemical reactions lose accuracy for intermediate values of the domain. Similarly, in Fig. 4.4(b), the perturbation solutions are shown to agree nicely with the numerical solutions for large values of Sc. The agreement between the numerical and perturbation solutions breaks down near $Sc = 1$.

Among these two solutions is very important to know which solution is physically relevant. It depends on the stability analysis of the solutions. Although two

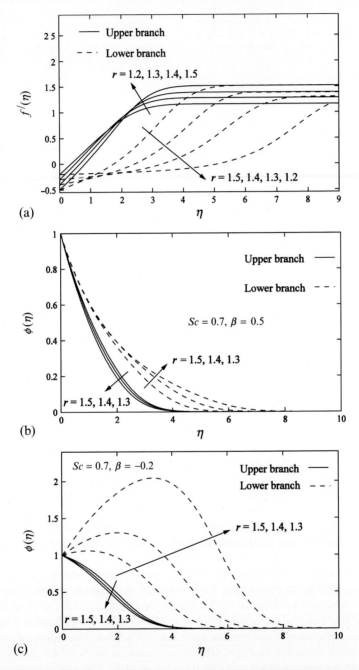

Figure 4.3 Effects of velocity ratio parameter r on (a) velocity profiles (b) concentration profiles in presence of destructive chemical reactions (c) concentration profiles in presence of constructive chemical reactions [16].

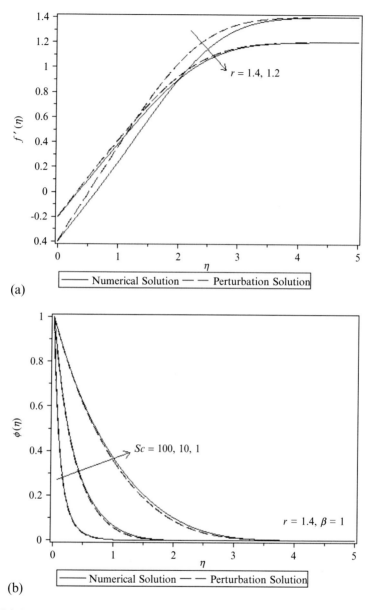

Figure 4.4 Comparison of perturbation and numerical solutions for (a) velocity profiles and (b) concentration profiles [16].

solution branches were present in the numerical solutions, only the upper solution branch was obtained analytically. This seems to suggest that the upper branch is the physically relevant solution branch. Furthermore, this assumption is supported by the fact that the upper branch solution is the only solution valid for $0 \leq r - 1 \leq 1$. In the recent literature for similar problems, it is shown that the lower solution branch is unstable and hence unphysical. Because the stability of the present problem is essentially an issue of the perturbed Blasius problem, we refer the reader to [22]. Note that in similar studies where two solution branches were obtained, one solution is found to be stable, whereas the extra solution branches are unstable [23–25].

4.3 Flow past a static vertical plate

During the last decades, flow of an incompressible viscous fluid and heat transfer phenomena over a static plate have received great attention owing to the abundance of practical applications in chemical and manufacturing processes. The interest in mixed convection boundary-layer flows of viscous incompressible fluids is increasing substantially because of its large number of practical applications in industry and manufacturing processes (Mukhopadhyay [26]). When the flow arises naturally simply owing to the effect of a density difference, resulting from a temperature or concentration difference in a body force field, such as the gravitational field, the process is termed free convection. The density difference gives rise to a buoyancy effect that generates the flow. The cooling of the heated body in ambient air generates such a flow in the region surrounding it. The buoyancy forces arising from the simultaneous effects of temperature differences play a significant role in mixed convective thermal diffusion when the flow velocity is relatively small and the temperature difference is relatively large (Subhashini et al. [27]). Cao and Baker [28] studied the mixed convective flow and heat transfer from a vertical plate and obtained local nonsimilar solutions to the boundary-layer equations. Pal and Shivakumara [3] discussed the effects of sparsely packed porous medium on mixed convection flow and heat transfer from a vertical plate. A similarity solution of mixed convection slip flow was reported by Bhattacharyya et al. [29]. In this paper, the buoyancy-opposed flow was not considered. Recently, Patil et al. [30] analyzed the unsteady mixed convection flow from a moving plate in a parallel free stream. They reported the combined effects of thermal radiation and Newtonian heating on mixed convection flow and heat transfer. We present here a few results obtained by Mukhopadhyay et al. [31].

We consider two-dimensional mixed convective steady boundary-layer flow of an incompressible viscous liquid over a flat plate. In the analysis of flow in porous media, the differential equation governing the fluid motion is based on the Darcy model, which accounts for the drag (represented by the Darcy term) exerted by the porous media. The governing equations for boundary layer flow and heat transfer can be expressed as follows:

$$\frac{\partial u}{\partial x} + \frac{\partial v}{\partial y} = 0, \tag{4.40}$$

$$u\frac{\partial u}{\partial x}+v\frac{\partial u}{\partial y}=\nu\frac{\partial^2 u}{\partial y^2}-\frac{\nu}{k}(u-u_\infty)+g\beta^*(T-T_\infty),\tag{4.41}$$

$$u\frac{\partial T}{\partial x}+v\frac{\partial T}{\partial y}=\frac{\kappa}{\rho c_p}\frac{\partial^2 T}{\partial y^2}+\frac{Q}{\rho c_p}(T-T_\infty).\tag{4.42}$$

Because the velocity of the fluid is low, the viscous dissipative heat is assumed to be negligible. The x coordinate is measured from the leading edge of the plate, and the y coordinate is measured along the normal to the plate. Here u and v are the components of velocity in the x and y directions, respectively; $\nu=\mu/\rho$ is the kinematic viscosity; μ is the coefficient of fluid viscosity; ρ is the fluid density; $k=k_0 x$ is the Darcy permeability of the porous medium; k_0 is the initial permeability; β^* is the volumetric coefficient of thermal expansion; T is the temperature; κ is the thermal conductivity of the fluid; u_∞ is the free stream velocity; T_∞ is the ambient temperature; and c_p is the specific heat at constant pressure. The term $Q=\frac{Q_0}{x}$ represents the moving heat source when $Q>0$ and the moving heat sink when $Q<0$.

The appropriate boundary conditions for the problem are given by (Bhattacharyya et al. [29,32])

$$u=L_1\frac{\partial u}{\partial y}, v=0, T=T_w+D_1\frac{\partial T}{\partial y}\quad\text{at }y=0,\tag{4.43}$$

$$u=u_\infty, T=T_\infty\quad\text{as }y\to\infty.\tag{4.44}$$

Here $T_w=T_\infty+\frac{T_0}{x}$ is the variable temperature of the plate; T_0 is a constant that measures the rate of increase of temperature along the plate; T_∞ is the free stream temperature assumed constant with $T_w>T_\infty$; $L_1=L(Re_x)^{1/2}$ is the velocity slip factor; $D_1=D(Re_x)^{1/2}$ is the thermal slip factor, with L and D being the initial values of velocity and thermal slip factors having the same dimension of length; and $Re_x=\frac{u_\infty x}{\nu}$ is the local Reynolds number.

We now introduce the following relations for u,v,

$$u=\frac{\partial\psi}{\partial y}, v=-\frac{\partial\psi}{\partial x},\tag{4.45}$$

where ψ is the stream function.

We also introduce the following dimensionless variables

$$\eta=y\sqrt{\frac{u_\infty}{\nu x}}, \psi=\sqrt{u_\infty\nu x}f(\eta)\quad\text{and}\quad\theta=\frac{T-T_\infty}{T_w-T_\infty}.\tag{4.46}$$

Using the above relations in equations (4.41) and (4.42), we obtain

$$\frac{\partial\psi}{\partial y}\frac{\partial^2\psi}{\partial x\partial y}-\frac{\partial\psi}{\partial x}\frac{\partial^2\psi}{\partial y^2}=\nu\frac{\partial^3\psi}{\partial y^3}-\frac{\nu}{k}\left(\frac{\partial\psi}{\partial y}-u_\infty\right)+g\beta^*(T_w-T_\infty)\theta,\tag{4.47}$$

and

$$\frac{\partial \psi}{\partial y}\left(\frac{\partial \theta}{\partial x}-\frac{\theta}{x}\right)-\frac{\partial \psi}{\partial x}\frac{\partial \theta}{\partial y}=\frac{\kappa}{\rho c_p}\frac{\partial^2 \theta}{\partial y^2}+\frac{Q}{\rho c_p}\theta. \tag{4.48}$$

The boundary conditions (4.43) and (4.44) then become

$$\frac{\partial \psi}{\partial y}=L_1\frac{\partial^2 \psi}{\partial y^2},\frac{\partial \psi}{\partial x}=0,\theta=1+D\frac{u_\infty}{\nu}\theta' \quad \text{at} \quad y=0, \tag{4.49}$$

$$\frac{\partial \psi}{\partial y}=u_\infty,\theta=0 \quad \text{as} \quad y\to\infty. \tag{4.50}$$

Using equation (4.46), the equations (4.47) and (4.48) finally can be put in the form

$$f''' +\frac{1}{2}ff'' -k_1\left(f'-1\right)+\lambda\theta=0, \tag{4.51}$$

$$\frac{1}{Pr}\theta'' +\frac{1}{2}f\theta' +f'\theta+Hs\theta=0, \tag{4.52}$$

where $k_1 =\dfrac{1}{Da_x Re_x}=\dfrac{\nu}{k_0 u_\infty}$ is the parameter of the porous medium (Damesh [14]),

$Da_x =\dfrac{k}{x^2}=\dfrac{k_0}{x}$ is the local Darcy number, $\lambda=g\beta^*\dfrac{T_0}{u_\infty^2}$ is the mixed convection param-

eter, $Pr =\dfrac{\mu c_p}{\kappa}$ is the Prandtl number, and $Hs=\dfrac{Q_0}{u_\infty \rho c_p}$ is the heat source ($Hs>0$) or

sink ($Hs<0$) parameter. Here k_1^{-1} reflects the effect of Darcian flows in the present problem.

For $\lambda>0$, the buoyancy forces act in the direction of the mainstream and fluid is accelerated in the manner of a favorable pressure gradient (assisting flow). On the other hand when $\lambda < 0$, the buoyancy forces oppose the motion, retarding the fluid in the boundary layer and acting as an adverse pressure gradient (opposing flow). Here $\lambda=0$ gives a purely forced convection situation.

The boundary conditions finally become

$$f' (\eta) = \delta f'' (\eta), f(\eta)=0,\theta(\eta)=1+\beta\theta' (\eta) \quad \text{at} \quad \eta=0, \tag{4.53}$$

and

$$f' (\eta)= 1, \theta(\eta)=0 \quad \text{at} \quad \eta\to\infty, \tag{4.54}$$

where $\delta=L\dfrac{u_\infty}{\nu}$ is the velocity slip parameter and $\beta=D\dfrac{u_\infty}{\nu}$ is the thermal slip param-

eter. When $k_1=0 = \lambda$, the equation (4.51) reduces to the equation of boundary layer flow on a flat plate at zero incidence.

4.3.1 Results and discussion

Effects of the mixed convection parameter λ on velocity and temperature are exhibited in Fig. 4.5(a) and (b). Fluid velocity is found to increase (Fig. 4.5(a)) with increasing values of λ for both cases of slip and no-slip boundary conditions, but the temperature decreases (Fig. 4.5(b)) with increasing values of λ. Because of favorable buoyancy effects, the fluid velocity increases within the boundary layer for buoyancy-aided flow ($\lambda > 0$), whereas for buoyancy-opposed flow ($\lambda < 0$), the reverse situation is noted (Fig. 4.5(a)). Velocity overshoot is noted for $\lambda = 0.3$ (Fig. 4.5(a)). This fact is also reported by Bhattacharyya et al. [29]. With the increase in λ, the temperature field is suppressed and consequently the thermal boundary-layer thickness becomes thinner. As a result, the rate of heat transfer from the plate increases (Fig. 4.5(b)). Actually $\lambda > 0$ means heating of the fluid or cooling of the surface (assisting flow). Increase in λ can increase the effects of the temperature field in the velocity distribution, which causes the enhancement of the velocity due to enhanced convection current. Physically, in the process of cooling, the free convection currents are carried away from the plate to the free stream, and as the free stream is in the upward direction, the free currents induce the velocity to enhance. As a result, the boundary-layer thickness increases. For opposing flow $\lambda < 0$, reverse effects are noted. No overshoot in temperature field, as reported by Bhattacharyya et al. [29], is noted. This is due to the presence of porous media.

The skin friction coefficient decreases with the increase in both velocity and thermal slip parameters (Fig. 4.6(a)). The skin friction coefficient is maximum at the no-slip condition, which is similar to the observations of Cao and Baker [28]. Temperature gradient at the wall decreases with velocity slip δ but increases with thermal slip β (Fig. 4.6(b)); that is, the rate of heat transfer increases with the increase of velocity slip δ, but it decreases with thermal slip parameter β.

4.4 Flow past a moving vertical plate

The importance of mixed convective phenomenon is increasing day by day because of the enhanced concern in science and technology about buoyancy-induced motions in the atmosphere, the bodies in water and quasi-solid bodies such as earth. Mixed convective boundary-layer flow of an electrically conducting fluid in the presence of a magnetic field has been the subject of a great number of investigations because of its fundamental importance in industrial and technological applications, including surface coating of metals, crystal growth, and reactor cooling. The Lorentz force is active and interacts with the buoyancy force in governing the flow and temperature fields. The effect of the Lorentz force is known to reduce the fluid velocities and suppress convection currents, and an external magnetic field is applied as a control mechanism in the material manufacturing industry. The MHD parameter is one of the important parameters by which the cooling rate can be controlled and a product of the desired quality can be achieved.

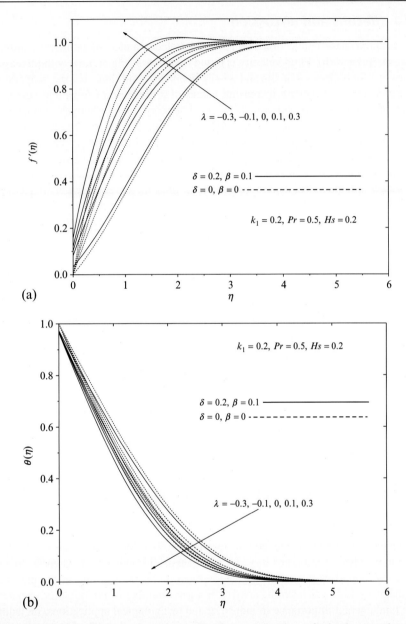

Figure 4.5 (a) Velocity (b) temperature profiles for several values of mixed convection parameter λ [31].

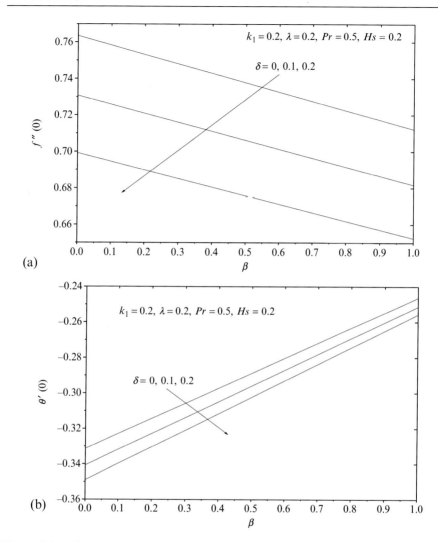

Figure 4.6 Variations of (a) skin-friction (b) heat transfer coefficients with velocity and thermal slip parameters [31].

4.4.1 Physical background

Free-convection boundary layer flow along a vertical heated plate in the presence of a magnetic field was investigated by Gupta [33], Emery [34], and Takhar [35]. Cheng and Minkowycz [36] and Cheng [37] studied free convective flow in a saturated porous medium. Wilks [38] had studied the combined forced and free convection flow along a semi-infinite plate extending vertically upwards. Boutros et al. [39] solved the steady free convective boundary-layer flow on a nonisothermal vertical plate. Recently, many studies were made on steady free convective boundary-layer flow on moving vertical plates considering the effect of buoyancy forces on the boundary layer (Chen and Strobel [40], Ramachandran et al. [41], Lee and Tsai [42]). The results obtained by Mukhopadhyay et al. [43] will be presented here.

Consider a combined convective, laminar boundary-layer flow of an electrically conducting incompressible, viscous liquid over a porous plate moving with nonuniform velocity $U(x)$ in the presence of a magnetic field $B(x)$. The governing equations of such type of flow are, in the usual notation,

$$\frac{\partial u}{\partial x} + \frac{\partial v}{\partial y} = 0, \tag{4.55}$$

$$u\frac{\partial u}{\partial x} + v\frac{\partial u}{\partial y} = \nu\frac{\partial^2 u}{\partial y^2} - \frac{\sigma B^2}{\rho}u + g\beta_1(T - T_\infty), \tag{4.56}$$

$$u\frac{\partial T}{\partial x} + v\frac{\partial T}{\partial y} = \frac{\kappa}{\rho c_p}\frac{\partial^2 T}{\partial y^2} - \frac{1}{\rho c_p}\frac{\partial q_r}{\partial y}. \tag{4.57}$$

Because the velocity of the fluid is low, the viscous dissipative heat is assumed to be negligible and the induced magnetic term is neglected here. The induced magnetic field is very small in comparison to the imposed magnetic field. The electric field is zero because no external electric field is applied, and the effect of polarization of the ionized fluid may be expected to be small and negligible. Here u and v are the components of velocity, respectively, in the x and y directions; μ is the coefficient of fluid viscosity; ρ is the fluid density; $\nu = \mu/\rho$ is the kinematic viscosity; T is the temperature; κ is the thermal conductivity of the fluid; σ is the electrical conductivity; β_1 is the volumetric coefficient of thermal expansion; g is the gravity field; T_∞ is the free stream temperature; q_r is the radiative heat flux; and c_p is the specific heat.

Using Rosseland approximation for radiation (Brewster [13]), we can write $q_r = -\frac{4\sigma_1}{3k^*}\frac{\partial T^4}{\partial y}$, where σ_1 is the Stefan-Boltzmann constant and k^* is the absorption coefficient.

Assuming that the temperature difference within the flow is such that T^4 may be expanded in a Taylor series, and expanding T^4 about T_∞ and neglecting higher orders, we get $T^4 \equiv 4T_\infty^3 T - 3T_\infty^4$.

Therefore, equation (4.57) becomes

$$u\frac{\partial T}{\partial x} + v\frac{\partial T}{\partial y} = \frac{\kappa}{\rho c_p}\frac{\partial^2 T}{\partial y^2} + \frac{16\sigma_1 T_\infty^3}{3\rho c_p k^*}\frac{\partial^2 T}{\partial y^2}. \tag{4.58}$$

The appropriate boundary conditions for the problem are given by

$$u = U(x), v = -V(x), T = T_w \text{ at } y = 0 \tag{4.59a}$$

$$u \to 0, T \to T_\infty \quad \text{as} \quad y \to \infty. \tag{4.59b}$$

Here T_w is the prescribed wall temperature, T_∞ is the free stream temperature assumed constant with $T_w > T_\infty$, $V(x)$ is the velocity of suction [$V(x) > 0$] or injection [$V(x) < 0$] of the conducting fluid.

We now introduce the following relations for u, v, and θ as

$$u = \frac{\partial \psi}{\partial y}, v = -\frac{\partial \psi}{\partial x} \text{ and } \theta = \frac{T - T_\infty}{T_w - T_\infty}, \tag{4.60}$$

where ψ is the stream function. Special types of velocity and temperature distributions at the wall are taken as

$$U(x) = ax^m, V(x) = V_0 x^{\frac{m-1}{2}}, T_w(x) = T_\infty + cx^n. \tag{4.61}$$

We also consider a special type of magnetic field (Chiam [44]):

$$B(x) = B_0 x^{\frac{m-1}{2}}.$$

Here a (>0), c (>0) are constants, B_0 is constant, and the power-law exponents m, n are also constants.

Using the equation (4.60) in the boundary-layer equation (4.56) and in the energy equation (4.58), we get the equations

$$\frac{\partial \psi}{\partial y} \frac{\partial^2 \psi}{\partial x \partial y} - \frac{\partial \psi}{\partial x} \frac{\partial^2 \psi}{\partial y^2} = g\beta_1 cx^n \theta + v \frac{\partial^3 \psi}{\partial y^3} - \frac{\sigma B_0^2}{\rho} x^{m-1} \frac{\partial \psi}{\partial y}, \tag{4.62}$$

and

$$n\theta \frac{\partial \psi}{\partial y} + x \frac{\partial \psi}{\partial y} \frac{\partial \theta}{\partial x} - x \frac{\partial \psi}{\partial x} \frac{\partial \theta}{\partial y} = x \left(\frac{\kappa}{\rho c_p} + \frac{16\sigma_1 T_\infty^3}{3\rho c_p k^*} \right) \frac{\partial^2 \theta}{\partial y^2}. \tag{4.63}$$

The boundary conditions given by equations (4.59a) and (4.59b) then become

$$\frac{\partial \psi}{\partial y} = ax^m, \frac{\partial \psi}{\partial x} = V_0 x^{\frac{m-1}{2}}, \theta = 1 \quad \text{at} \quad y = 0, \tag{4.64a}$$

$$\frac{\partial \psi}{\partial y} \to 0, \theta \to 0 \quad \text{as} \quad y \to \infty. \tag{4.64b}$$

We now introduce the simplified form of Lie-group transformations, namely, the scaling group of transformations (Mukhopadhyay et al. [45])

$$\Gamma : \left\{ \begin{array}{l} x^* = xe^{\varepsilon \alpha_1}, y^* = ye^{\varepsilon \alpha_2}, \psi^* = \psi e^{\varepsilon \alpha_3}, \\ u^* = ue^{\varepsilon \alpha_4}, v^* = ve^{\varepsilon \alpha_5}, \theta^* = \theta e^{\varepsilon \alpha_6}. \end{array} \right\} \tag{4.65}$$

Substituting equations (4.65) into (4.62) and (4.63), we get

$$e^{\varepsilon(\alpha_1+2\alpha_2-2\alpha_3)}\left(\frac{\partial\psi^*}{\partial y^*}\frac{\partial^2\psi^*}{\partial x^*\partial y^*}-\frac{\partial\psi^*}{\partial x^*}\frac{\partial^2\psi^*}{\partial y^{*2}}\right)=g\beta_1 ce^{-\varepsilon(n\alpha_1+\alpha_6)}x^{*n}\theta^*$$

$$+\nu e^{\varepsilon(3\alpha_2-\alpha_3)}\frac{\partial^3\psi^*}{\partial y^{*3}}-\frac{\sigma B_0^2}{\rho}e^{\varepsilon(\alpha_2-\alpha_3-m\alpha_1+\alpha_1)}x^{*(m-1)}\frac{\partial\psi^*}{\partial y^*},\tag{4.66}$$

$$n\theta^*\frac{\partial\psi^*}{\partial y^*}e^{\varepsilon(\alpha_2-\alpha_3-\alpha_6)}+\left(x^*\frac{\partial\psi^*}{\partial y^*}\frac{\partial\theta^*}{\partial x^*}-x^*\frac{\partial\psi^*}{\partial x^*}\frac{\partial\theta^*}{\partial y^*}\right)e^{\varepsilon(\alpha_1+\alpha_2-\alpha_3-\alpha_6-\alpha_1)}$$

$$=x^*\left(\frac{\kappa}{\rho c_p}+\frac{16\sigma_1 T_\infty^3}{3\rho c_p k^*}\right)\frac{\partial^2\theta^*}{\partial y^{*2}}e^{\varepsilon(2\alpha_2-\alpha_1-\alpha_6)}.\tag{4.67}$$

The system will remain invariant under the group of transformations Γ, and we have the following relations among the parameters:

$$\alpha_1+2\alpha_2-2\alpha_3=-n\alpha_1-\alpha_6=3\alpha_2-\alpha_3=\alpha_2-\alpha_3-m\alpha_1+\alpha_1$$

and

$$\alpha_2-\alpha_3-\alpha_6=2\alpha_2-\alpha_1-\alpha_6..$$

Solving the above equations, we get

$$(m+1)\alpha_1=2\alpha_3,(m-1)\alpha_1=-2\alpha_2,\text{ and }(2m-n-1)\alpha_1=\alpha_6.$$

In view of these, the boundary conditions become

$$\frac{\partial\psi^*}{\partial y^*}=ax^{*m},\frac{\partial\psi^*}{\partial x^*}=V_0 x^{*\left(\frac{m-1}{2}\right)},\theta^*=1\ \text{ at }\ y^*=0\tag{4.68a}$$

and

$$\frac{\partial\psi^*}{\partial y^*}\to 0,\theta^*\to 0\ \text{ as }\ y^*\to\infty\tag{4.68b}$$

with the additional conditions $\alpha_6=0$, which gives $n=(2m-1)$.

Thus Γ reduces to a one-parameter group of transformations:

$$x^*=xe^{\varepsilon\alpha_1},y^*=ye^{-\varepsilon\left(\frac{m-1}{2}\right)\alpha_1},\psi^*=\psi e^{\varepsilon\left(\frac{m+1}{2}\right)\alpha_1},$$

$$u^*=ue^{\varepsilon m\alpha_1},v^*=ve^{\varepsilon\left(\frac{m-1}{2}\right)\alpha_1},\theta^*=\theta.\tag{4.69}$$

Expanding by Taylor's method in powers of ε and keeping terms up to the order ε, we get from equation (4.69)

$$x^* - x = x\varepsilon\alpha_1, y^* - y = -y\varepsilon\left(\frac{m-1}{2}\right)\alpha_1, \psi^* - \psi = \psi\varepsilon\left(\frac{m+1}{2}\right)\alpha_1,$$

$$u^* - u = u\varepsilon m\alpha_1, v^* - v = v\varepsilon\left(\frac{m-1}{2}\right)\alpha_1, \theta^* - \theta = 0. \tag{4.70}$$

In terms of differentials, these yield

$$\frac{dx}{\alpha_1 x} = \frac{dy}{-\alpha_1\dfrac{m-1}{2}y} = \frac{d\psi}{\alpha_1\dfrac{m+1}{2}\psi} = \frac{du}{\alpha_1 m u} = \frac{dv}{\alpha_1 v\dfrac{m-1}{2}} = \frac{d\theta}{0}. \tag{4.71}$$

Solving the above equations, we get $yx^{(m-1)/2} = \eta, \psi = x^{(m+1)/2}F(\eta), \theta = \theta(\eta)$. In view of the above relations, the equations (4.66) and (4.67) become

$$mF'^2 - \left(\frac{m+1}{2}\right)FF'' = (g\beta_1 c)\theta + \nu F''' - aM^2 F', \tag{4.72}$$

$$(2m-1)\theta F' - \left(\frac{m+1}{2}\right)F\theta' - \left(\frac{\kappa}{\rho c_p} + \frac{16\sigma_1 T_\infty^3}{3\rho c_p k^*}\right)\theta'' = 0. \tag{4.73}$$

The boundary conditions are transformed to

$$F'(\eta) = a, F(\eta) = \frac{2V_0}{m+1} \text{ and } \theta(\eta) = 1 \text{ at } \eta = 0, \tag{4.74a}$$

$$F'(\eta) \to 0, \theta(\eta) \to 0 \text{ as } \eta \to \infty. \tag{4.74b}$$

Again, we introduce the following transformations for η, F, and θ in equations (4.72) and (4.73):

$$\eta = (g\beta_1 c)^{a_1} \nu^{b_1} a^{c_1} \eta^*, F = (g\beta_1 c)^{a_1'} \nu^{b_1'} a^{c_1'} F^*,$$

$$\theta = (g\beta_1 c)^{a_1''} \nu^{b_1''} a^{c_1''} \bar{\theta}$$

and obtain $a_1' = a_1 = 0, a_1'' = -1, b_1' = b_1 = \frac{1}{2}, b_1'' = 0, c_1 = -\frac{1}{2}, c_1' = \frac{1}{2}, c_1'' = 2$.

Taking $F^* = f$ and $\bar{\theta} = \theta$, the equations (4.72) and (4.73) finally take the following form:

$$f''' + \frac{m+1}{2}ff'' - mf'^2 + \theta - M^2 f' = 0, \tag{4.75}$$

and

$$\frac{3N+4}{3NPr}\theta'' + \frac{m+1}{2}f\theta' - (2m-1)\theta f' = 0, \tag{4.76}$$

where $\sigma B_0^2/\rho = aM^2$, M is the Hartmann number, $Pr = \mu c_p/\kappa$ is the Prandtl number, and $N = \kappa k^*/(4\sigma_1 T_\infty^3)$ is the radiation parameter.

The boundary conditions take the forms

$$f' = 1, f = S, \theta = 1 \quad \text{at} \quad \eta^* = 0 \tag{4.77a}$$

and

$$f' \to 0, \theta \to 0 \quad \text{as} \quad \eta^* \to \infty, \tag{4.77b}$$

where $S = \dfrac{2V_0}{(m+1)(av)^{1/2}}$ is the suction ($V_0 > 0$)/injection($V_0 < 0$) parameter.

4.4.2 Results and discussion

To get a clear insight of the physical problem, numerical computations are carried out for numerous values of the governing parameters, and the numerical values are illustrated through figures. Figure 4.7 shows the streamwise velocity profile for several values of m in the absence of magnetic field and when there is no suction/injection. The velocity profiles have maximum values within the boundary layer and away from the surface. Initially, streamwise velocity is found to decrease with the increasing m but the opposite trend is noted after a certain distance from the surface (Fig. 4.7).

Now the result for the variation of the magnetic parameter M is presented in Fig. 4.8. In this figure, streamwise velocity profiles are shown for different values of M ($M = 0$, 0.5, 1) with $m = 0.5$, $N = 1$, $Pr = 0.5$, in case of constant suction

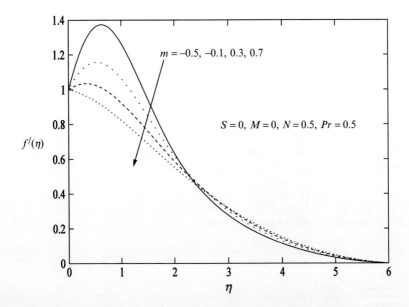

Figure 4.7 Velocity profiles for variable values of power law index m [43].

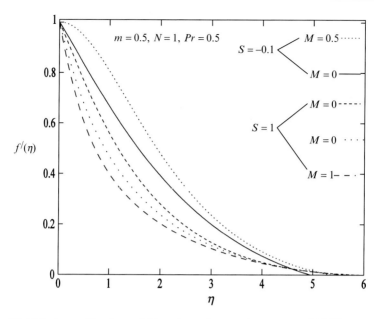

Figure 4.8 Velocity profiles for variable values of magnetic parameter M [43].

($S = 1$) and in case of constant blowing ($S = -0.1$). The streamwise velocity curves show that the rate of transport decreases with the increase in M in case of suction. It clearly indicates that the transverse magnetic field opposes the transport phenomena. This is due to the fact that variation of M leads to the variation of the Lorentz force due to the magnetic field, and the Lorentz force produces more resistance to transport phenomena. In all cases, the velocity vanishes at some large distance from the plate (Fig. 4.8).

4.5   Nanofluid boundary layers over a moving plate

The forced convection over a flat plate/sheet has been widely studied over the past few decades. Earlier investigators were mainly interested to find the similarity solutions for the boundary-layer flow problems. In fluid mechanics, the problem of viscous boundary-layer flow on a moving or fixed flat plate is a classical problem. Flow and heat transfer of a viscous fluid over a moving surface has many important applications in the modern industry, namely, polymer industry, glass fiber drawing, crystal growing, plastic extrusion, continuous casting, etc. (Magyari and Keller [46]). Siekman [47], Klemp and Acrivos [48], Abdulhafez [49], Chappidi and Gunnerson [50], Hussaini et al. [51], Lin and Haung [52], and Sparrow and Abraham [53] reported the flow and heat transfer characteristics for moving wall laminar boundary-layer problems. Cortell [17] extended the work of Afzal et al. [54] for constant as well as prescribed power-law surface temperature. However, the existence of dual solutions was not discussed in that study. Recently, Ishak et al. [18] showed that dual solutions exist when the velocity ratio exceeds unity, that is, when the sheet moves in opposite direction to the free stream. The effects of suction and injection on the flow and thermal fields for constant surface temperature were also reported in that study. But in his study the prescribed surface temperature was disregarded. Of late, Mukhopadhyay [55] investigated the case of prescribed surface temperature of the second degree and reported the existence of dual solutions.

4.5.1 Physical background

The nanofluids have many industrial applications because materials of nanometer size have unique physical and chemical properties. Nanofluids have attracted great interest recently because of their greatly enhanced thermal properties. This phenomenon suggests the possibility of using nanofluids in advanced nuclear systems. Suspensions of metal nanoparticles are also being developed for other purposes, such as medical applications, including cancer therapy.

Nanoparticles in the base fluid (nanofluid) provide many advantages in a solar energy absorption system. The interdisciplinary nature of nanofluid research presents a great opportunity for exploration and discovery at the frontiers of nanotechnology. In this section, we shall report the results obtained by Mandal et al. [56], who studied the heat transfer for boundary-layer forced convective flow of a nanofluid past a moving flat surface parallel to a moving stream.

We consider a forced convective, two-dimensional steady laminar boundary-layer flow of a nanofluid over a flat surface moving with constant velocity U_w in the same or opposite direction to the free stream U_∞ (directed toward the positive x-direction). The x-axis extends parallel to the surface, whereas the y-axis extends upwards, normal to the surface. The governing equations for such boundary-layer flows of nanofluid are written as

$$\frac{\partial u}{\partial x} + \frac{\partial v}{\partial y} = 0, \tag{4.78}$$

$$u\frac{\partial u}{\partial x} + v\frac{\partial u}{\partial y} = \nu\frac{\partial^2 u}{\partial y^2}, \tag{4.79}$$

$$u\frac{\partial T}{\partial x} + v\frac{\partial T}{\partial y} = \alpha_m\frac{\partial^2 T}{\partial y^2} + \tau\left[D_B\frac{\partial C}{\partial y}\frac{\partial T}{\partial y} + \frac{D_T}{T_\infty}\left(\frac{\partial T}{\partial y}\right)^2\right], \tag{4.80}$$

$$u\frac{\partial C}{\partial x} + v\frac{\partial C}{\partial y} = D_B\frac{\partial^2 C}{\partial y^2} + \frac{D_T}{T_\infty}\frac{\partial^2 T}{\partial y^2}. \tag{4.81}$$

The appropriate boundary conditions for the problem are given by

$$u = U_w, v = 0, T = T_w, C = C_w \quad \text{at} \quad y = 0, \tag{4.82a}$$

$$u \to U_\infty, T \to T_\infty, C \to C_\infty, \text{ as } y \to \infty. \tag{4.82b}$$

Here $T_w = T_\infty + Ax^n$ is the prescribed surface temperature and $C_w = C_\infty + Bx^n$ is the prescribed concentration and n is the power law exponent.

With the help of a composite velocity $U = U_w + U_\infty$, we now introduce the dimensionless variables

$$\eta = y\sqrt{\frac{U}{2\nu x}}, \quad u = Uf'(\eta), \quad v = U\frac{\eta f'(\eta) - f(\eta)}{\sqrt{2Re_x}}, \tag{4.83}$$

$$\theta - \frac{T - T_\infty}{T_w - T_\infty}, \phi = \frac{C - C_\infty}{C_w - C_\infty}. \tag{4.84}$$

Here $Re_x = \dfrac{Ux}{\nu}$ is the local Reynolds number, $\nu = \dfrac{\mu}{\rho_f}$ is the kinematic viscosity of the fluid, μ is the viscosity of the fluid, ρ_f is the density of the base fluid, α_m is the thermal diffusivity, D_B is the Brownian diffusion coefficient, D_T is the thermophoresis diffusion coefficient, c_p is the specific heat at constant pressure, and τ is the ratio of the effective heat capacity of the nanoparticle material and the heat capacity of the ordinary fluid.

Using the relations (4.83) and (4.84) in the governing boundary-layer equations (4.79)–(4.81), we get the equations

$$f''' + ff'' = 0, \tag{4.85}$$

$$\theta'' + Pr\left(f\theta' - 2nf'\theta\right) + Nb\,\theta'\,\phi' + Nt\left(\theta'\right)^2 = 0, \tag{4.86}$$

$$\phi'' + Sc\left(f\phi' - 2nf'\phi\right) + \frac{Nt}{Nb}\theta'' = 0. \tag{4.87}$$

The transformed boundary conditions then become

$$f' = 1 - r, f = 0, \theta = 1, \phi = 1 \text{ at } \eta = 0,$$

and

$$f' = r, \theta = 0, \phi \to 0 \text{ as } \eta \to \infty, \tag{4.88}$$

where $r = \dfrac{U_\infty}{U}$ is the velocity ratio parameter and where $Nt = \dfrac{\tau D_T(T_w - T_\infty)}{T_\infty \nu}$, $Nb = \dfrac{\tau D_B(C_w - C_\infty)}{\nu}$, $Pr = \dfrac{\nu}{\alpha_m}$, and $Sc = \dfrac{\nu}{D_B}$ are the thermophoresis parameter, Brownian motion parameter, Prandtl number, and Schmidt number, respectively.

Equations (4.85)–(4.87) along with boundary conditions (4.88) were solved numerically by the shooting method. The details about the directions of the moving wall and the free stream can be found from Mukhopadhyay [55].

4.5.2 Results and discussion

To analyze the results, numerical computation has been carried out and the numerical values are presented through tables and figures. Tables 4.1–4.3 display results for wall values for the gradients of temperature and concentration functions that are proportional to the Nusselt number and Sherwood number, respectively. Table 4.1 shows that as the Brownian motion parameter Nb increases, the surface mass transfer rates increase, whereas the surface heat transfer rate decreases. Brownian motion decelerates the flow in the nanofluid boundary layer. Brownian diffusion promotes heat conduction. The nanoparticles increase the surface area for heat transfer. Nanofluid is a two-phase fluid where the nanoparticles move randomly and increase the energy exchange rates. Brownian motion reduces nanoparticle diffusion. Figure 4.9(a) and (b) shows that as Nb increases, temperature increases but concentration decreases within the boundary layer.

Table 4.2 shows that as Nt increases, the heat transfer rate decreases for both solution branches, and the mass transfer rate decreases initially for both the branches. The thermophoresis parameter, Nt, appears in the thermal and concentration boundary-layer equations. As we note, it is coupled with the temperature function and plays a strong role in determining the diffusion of heat and nanoparticle concentration in the boundary layer. From Fig. 4.10(a) and 4.10(b), we note that the temperature and nanoparticle concentration are elevated as Nt increases.

As Le increases, the heat transfer rate decreases whereas the mass transfer rate increases as shown by Table 4.3. Le represents the ratio of molecular thermal diffusivity to mass diffusivity (Sc and Pr). As Le increases, the thermal diffusivity is more pronounced than the mass diffusivity. Because large values of Le make the molecular diffusivity smaller, concentration decreases as Le increases.

The effective thermal conductivity of the nanofluid increases because of the Brownian motion and thermophoresis of nanoparticles. Both Brownian diffusion and thermophoresis give rise to cross-diffusion terms that are similar to the familiar Soret and Dufour cross-diffusion terms that arise with a binary fluid discussed by Lakshmi Narayana et al. [57].

Table 4.1 **Values of $f''(0), -\theta'(0), -\phi'(0)$ for variable Nb when $Nt = 0.3$, $n = 0.5$, $Sc = 10$, $Pr = 10$ [56]**

	$f''(0)$		$-\theta'(0)$		$-\phi'(0)$	
Nb	Upper branch	Lower branch	Upper branch	Lower branch	Upper branch	Lower branch
0.1	0.533707	0.001491	1.396806	0.869207	1.047841	0.881666
0.2			1.330423	0.825062	1.428545	1.011311
0.3			1.266499	0.782562	1.553021	1.052862
0.4			1.204970	0.741746	1.613487	1.072464
0.5			1.145809	0.702389	1.648375	1.083226

Table 4.2 **Values of** $f''(0), -\theta'(0), -\phi'(0)$ **for variable** Nt **when** $Nb = 0.3$, $n = 0.5$, $Sc = 10$, $Pr = 10$ [56]

	$f''(0)$		$-\theta'(0)$		$-\phi'(0)$	
Nt	Upper branch	Lower branch	Upper branch	Lower branch	Upper branch	Lower branch
0.0	0.533707	0.001491	1.421437	0.885430	1.641448	1.032086
0.1			1.367715	0.849738	1.584473	1.021343
0.2			1.316098	0.815501	1.555633	1.028707
0.3			1.266499	0.782562	1.553021	1.052862
0.4			1.218879	0.751151	1.574679	1.092509
0.5			1.173176	0.720987	1.618820	1.146365

Table 4.3 **Values of** $-\theta'(0), -\phi'(0)$ **for variable** Sc **when** $Nb = 0.3$, $Nt = 0.3$, $n = 0.5$, $Pr = 10$ [56]

	$-\theta'(0)$		$-\phi'(0)$	
Sc	Upper branch	Lower branch	Upper branch	Lower branch
1	1.336763	0.872467	0.791372	0.248509
10	1.266499	0.782562	1.553021	1.052862
100	1.166775	0.747938	2.812732	1.180894

4.6 Unsteady boundary-layer flow caused by an impulsively stretching plate

The study of laminar flow over a stretching plate in a viscous fluid is of considerable interest because of its important bearings on several technological processes. Boundary-layer flows of an incompressible fluid over a stretching plate have many important applications in engineering. The aerodynamic extrusion of plastic sheets, the boundary layer along a liquid film condensation process, the cooling process of a metallic plate in a cooling bath, and in the glass and polymer industries are just to name a few. In the present section, we present the results of Liao [58] as discussed in Vajravelu and Van Gorder [59], where solutions for an unsteady boundary-layer flow due to an impulsively stretched plate were obtained using the homotopy analysis method.

Sakiadis [60], Crane [61], Banks [62], Grubka and Bobba [63], and Ali [64] investigated the flow past an impermeable plate whereas Erickson et al. [65], Gupta and Gupta [66], Chen and Char [67], Chaudhary et al. [68], Elbashbeshy [69], and Magyari and Keller [70] investigated the flow for the permeable plate. Some researchers

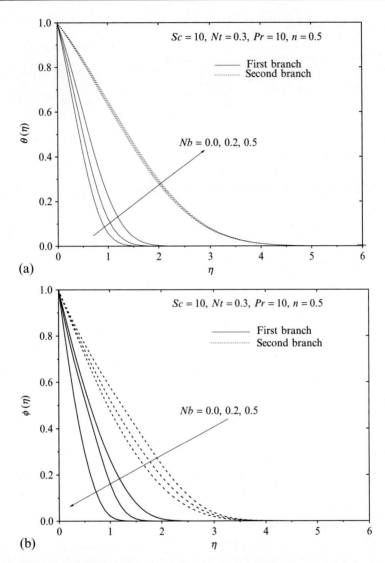

Figure 4.9 (a) Temperature and (b) concentration distribution for variable values of Brownian motion parameter Nb [56].

[71–76] considered the unsteady boundary layers due to an impulsively started flat plate. However, the unsteady boundary-layer flows due to an impulsively stretching plate in a viscous fluid [76–79] is addressed relatively little. Recently, the unsteady boundary-layer flow due to an impulsively stretching surface in a rotating fluid was solved by Nazar et al. [79] using a transformation found by Williams and Rhyne [80], and they obtained the numerical solution by applying the so-called Keller-box

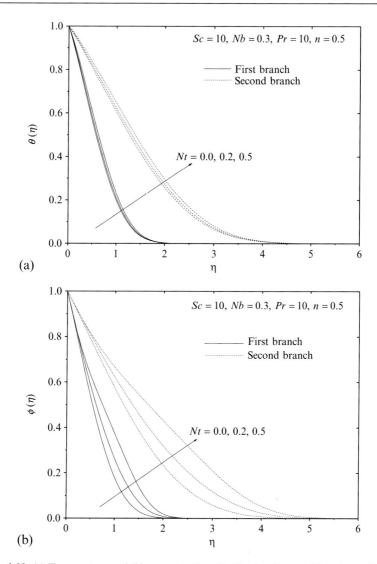

Figure 4.10 (a) Temperature and (b) concentration distribution for variable values of thermophoresis parameter Nt [56].

method. They also obtained a first-order perturbation approximation. It is very diffi-cult to obtain analytic solutions of unsteady boundary-layer flows that are valid for all time. Many researchers use the perturbation techniques, but the corresponding analytic solutions are valid only for a short time [74,76,79]. To the authors' knowl-edge, no one has reported any analytic solution of the unsteady boundary-layer flow over a semi-infinite flat plate, which is valid and accurate for all time. Currently, an analytical method for strongly nonlinear problems, namely, the homotopy analysis method [81], has been developed and successfully applied to many kinds of non-linear problems in science and engineering [82–91]. Here, the homotopy analysis method is employed to give an analytic solution of the unsteady boundary-layer flows caused by an impulsively stretching plate, which is valid and accurate for all time [59].

4.6.1 Mathematical description

Consider an unsteady boundary layer developed by an impulsively stretching plate in a constant pressure viscous flow, governed by (see, Vajravelu and Van Gorder [59])

$$\frac{\partial u}{\partial t} + u\frac{\partial u}{\partial x} + v\frac{\partial u}{\partial y} = v\frac{\partial^2 u}{\partial y^2}, \tag{4.89}$$

$$\frac{\partial u}{\partial x} + \frac{\partial v}{\partial y} = 0, \tag{4.90}$$

subject to the boundary conditions

$$t \geq 0: \ u = ax, \ v = 0 \ \text{ at } \ y = 0, \tag{4.91a}$$

$$u \rightarrow 0 \ \text{ as } \ y \rightarrow +\infty \tag{4.91b}$$

where $a > 0$, and the initial conditions

$$t = 0: u = v = 0 \ \text{ at all points}(x, y). \tag{4.91c}$$

Following Seshadri et al. [76] and Nazar et al. [79], we use Williams and Rhyne's similarity transformation [80]

$$\psi = \sqrt{av\xi}\, xf(\eta, \xi), \ \eta = \sqrt{\frac{a}{v\xi}}\, y, \ \xi = 1 - \exp(-\tau), \ \tau = at, \tag{4.92}$$

where ψ denotes the stream function.

For $\xi \geq 0$, the governing equations become

$$\frac{\partial^3 f}{\partial \eta^3} + \frac{1}{2}(1-\xi)\eta\frac{\partial^2 f}{\partial \eta^2} + \xi\left[f\frac{\partial^2 f}{\partial \eta^2} - \left(\frac{\partial f}{\partial \eta}\right)^2\right] = \xi(1-\xi)\frac{\partial^2 f}{\partial \eta \partial \xi}, \tag{4.93}$$

subject to the boundary conditions

$$f(0, \xi) = 0, \quad \left.\frac{\partial f}{\partial \eta}\right|_{\eta=0} = 1, \quad \left.\frac{\partial f}{\partial \eta}\right|_{\eta=+\infty} = 0. \tag{4.94}$$

For $\xi = 0$ (corresponding to $\tau = 0$), equation (4.93) becomes

$$\frac{\partial^3 f}{\partial \eta^3} + \frac{1}{2}\eta\frac{\partial^2 f}{\partial \eta^2} = 0, \tag{4.95}$$

which is the Rayleigh type of equation, subject to the conditions

$$f(0, 0) = 0, \quad \left.\frac{\partial f}{\partial \eta}\right|_{\eta=0,\xi=0}, \quad \left.\frac{\partial f}{\partial \eta}\right|_{\eta=+\infty,\xi=0} = 0. \tag{4.96}$$

The exact solution of the above equation is given by

$$f(\eta, 0) = \eta\,\mathrm{erfc}(\eta/2) + \frac{2}{\sqrt{\pi}}\left[1 - \exp\left(-\eta^2/4\right)\right], \tag{4.97}$$

where $\mathrm{erfc}(\eta) = \dfrac{2}{\sqrt{\pi}}\displaystyle\int\limits_{\eta}^{+\infty} \exp\left(-z^2\right) dz$ is the error function.

When $\xi = 1$, corresponding to $\tau \to +\infty$, we have from equation (4.93) that

$$\frac{\partial^3 f}{\partial \eta^3} + f\frac{\partial^2 f}{\partial \eta^2} - \left(\frac{\partial f}{\partial \eta}\right)^2 = 0, \tag{4.98}$$

subject to

$$f(0, 1) = 0, \quad \left.\frac{\partial f}{\partial \eta}\right|_{\eta=0,\xi=1} = 1, \quad \left.\frac{\partial f}{\partial \eta}\right|_{\eta=+\infty,\xi=0} = 0. \tag{4.99}$$

The exact solution of the above equation is

$$f(\eta, 1) = 1 - \exp(-\eta). \tag{4.100}$$

As n increases from 0 to 1, $f(\eta, \xi)$ varies from the initial solution (4.97) to the steady solution (4.100). It is to be noted that although $f'(+\infty, \xi) \to 0$ exponentially for all ξ, where the prime denotes the differentiation with respect to η, $f'(+\infty, 0)$ of the initial solution tends to 0 much more quickly than $f'(+\infty, 1)$ of the steady solution. For this, mathematically, the initial solution (4.97) is different in essence from the steady one (4.100). This might be the reason why it is so hard to give an accurate analytic solution uniformly valid for all time $0 \le \tau < +\infty$. When $\xi = 0$, we have

$$\left.\frac{\partial^2 f}{\partial \eta^2}\right|_{\eta=0,\xi=0} = -\frac{1}{\sqrt{\pi}} \tag{4.101}$$

and for $\xi = 1$,

$$\frac{\partial^2 f}{\partial \eta^2}\bigg|_{\eta=0,\,\xi=1} = -1 . \tag{4.102}$$

The skin friction coefficient is given by

$$c_f^x(x, \xi) = (\xi Re_x)^{-1/2} f''(0, \xi), \, 0 \leq \xi \leq 1, \tag{4.103}$$

where $Re_x = \dfrac{ax^2}{\nu}$ is the local Reynolds number.

4.6.2 Results and discussion

Homotopy analytic solution: In this section, we employ the homotopy analysis method described in [59] to solve equations (4.93) and (4.94). We should avoid the appearance of the error function and its powers so that high-order approximations can be obtained. From (4.94), (4.95), (4.97), and (4.100), it is reasonable to assume that $f(\eta, \xi)$ could be expressed by the following set of base functions

$$\left\{ \xi^k \eta^m \exp(-n\eta) \,|\, k \geq 0, \, m \geq 0, \, n \geq 0 \right\} \tag{4.104}$$

such that

$$f(\eta, \xi) = a_0^{0,0} + \sum_{k=0}^{+\infty} \sum_{m=0}^{+\infty} \sum_{n=1}^{+\infty} a_k^{m,n} \xi^k \eta^m \exp(-n\eta), \tag{4.105}$$

where $a_k^{m,n}$ is a coefficient. It provides us with the so-called Rule of Solution Expression (see [81,82]). From (4.93), (4.94), and (4.105), it is straightforward to choose the initial approximation

$$f_0(\eta, \xi) = 1 - \exp(-\eta), \tag{4.106}$$

which is exactly the same as the steady-state solution $f(\eta, 1)$, and the auxiliary linear operator

$$L[\phi(\eta, \xi; q)] = \frac{\partial^3 \phi}{\partial \eta^3} - \frac{\partial \phi}{\partial \eta}, \tag{4.107}$$

which has the property

$$L[C_1 + C_2 \exp(-\eta) + C_3 \exp(\eta)] = 0. \tag{4.108}$$

From (4.93), we define the nonlinear operator

$$N[\phi(\eta, \xi; q)] = \frac{\partial^3 \phi}{\partial \eta^3} + \frac{1}{2}(1 - \xi)\eta \frac{\partial^2 \phi}{\partial \eta^2}$$
$$+ \xi \left[\phi \frac{\partial^2 \phi}{\partial \eta^2} - \left(\frac{\partial \phi}{\partial \eta} \right)^2 \right] - \xi(1 - \xi) \frac{\partial^2 \phi}{\partial \eta \partial \xi}. \tag{4.109}$$

Let $\hbar$ denote a nonzero auxiliary parameter. We construct the so-called zero-order deformation equation (see [81,82])

$$(1 - q)L[\phi(\eta, \xi; q) - f_0(\eta, \xi)] = q\hbar N[\phi(\eta, \xi; q)], \tag{4.110}$$

subject to the boundary conditions

$$\phi(0, \xi; q) = 0, \quad \frac{\partial \phi(\eta, \xi; q)}{\partial \eta} \bigg|_{n=+\infty} = 1, \quad \frac{\partial \phi(\eta, \xi; q)}{\partial \eta} \bigg|_{n=+\infty} = 0, \tag{4.111}$$

where $q \in [0, 1]$ is an embedding parameter. Obviously, when $q = 0$ and $q = 1$, we have

$$\phi(\eta, \xi; 0) = f_0(\eta, \xi) \tag{4.412}$$

and

$$\phi(\eta, \xi; 1) = f(\eta, \xi), \tag{4.113}$$

respectively. Thus, as q increases from 0 to 1, $\phi(\eta,\xi;q)$ varies from the initial approximation $f_0(\eta,\xi)$ to the solution $f(\eta,\xi)$ of the original equations (4.93) and (4.94). Assume that the auxiliary parameter $\hbar$ is so properly chosen that the Taylor series of $\phi(\eta,\xi;q)$ expanded with respect to the embedding parameter, that is,

$$\phi(\eta, \xi; q) = \phi(\eta, \xi; 0) + \sum_{n=1}^{+\infty} f_n(\eta, \xi)q^n, \tag{4.114}$$

where

$$f_n(\eta, \xi) = \frac{1}{n!} \frac{\partial^n \phi(\eta, \xi; q)}{\partial q^n} \bigg|_{q=0}, \tag{4.115}$$

converges at $q = 1$. Then, we have that

$$f(\eta, \xi) = f_0(\eta, \xi) + \sum_{n=1}^{+\infty} f_n(\eta, \xi). \tag{4.116}$$

Write $\vec{f}_n = \{f_0, f_1, f_2, ..., f_n\}$. (Differentiating the zero-order deformation equations m times with respect to q, then dividing by $m!$, and finally setting $q - 0$, we have the mth-order deformation equations (see [81,82])

$$L[f_m(\eta, \xi) - x_m f_{m-1}(\eta, \xi)] = \hbar R_m\left(\vec{f}_{m-1}, \eta, \xi\right), \tag{4.117}$$

subject to the boundary conditions

$$f_m(0, \xi) = 0, \quad \frac{\partial f_m(\eta, \xi)}{\partial q}\bigg|_{n=0,} = 0, \quad \frac{\partial f_m(\eta, \xi)}{\partial q}\bigg|_{n=+\infty} = 0, \tag{4.118}$$

where

$$R_m\left(\vec{f}_{m-1}, \eta, \xi\right) = \frac{\partial^3 f_{m-1}}{\partial \eta^3} + \frac{1}{2}(1-\xi)\eta\frac{\partial^2 f_{m-1}}{\partial \eta^2} - \xi(1-\xi)\frac{\partial^2 f_{m-1}}{\partial \eta \partial \xi}$$
$$+ \xi\sum_{n=0}^{m-1}\left[f_{m-1-n}\frac{\partial^2 f_n}{\partial \eta^2} - \frac{\partial f_{m-1-n}}{\partial \eta}\frac{\partial f_n}{\partial \eta}\right] \tag{4.119}$$

and

$$\chi_n = \begin{cases} 1, & n > 1, \\ 0, & n = 1. \end{cases} \tag{4.120}$$

Let $f_m^*(\eta, \xi)$ denote a special solution of equation (4.117). From (4.108), its general solution reads

$$f_m(\eta, \xi) = f_m^*(\eta, \xi) + C_1 + C_2\exp(-\eta) + C_3\exp(\eta), \tag{4.121}$$

where the coefficients C_1, C_2, and C_3 are determined by the given boundary conditions. In this way, it is easy to solve the linear equations resulting from the higher-order deformation equations successively. Note that unlike the previous perturbation approach, the special function $\mathrm{erfc}(\eta/2)$ does not appear in the high-order deformation equations. So, we can easily obtain results at rather high-order of approximations, especially by means of the symbolic computation software such as Mathematica. In this way, we can obtain accurate analytic approximations uniformly valid for all time τ, as described below (see [59]).

It is important to ensure that the solution series is convergent. Note that the series contains an auxiliary parameter $\hbar$; obviously, the convergence of the series is determined by this auxiliary parameter. Because the initial approximation is exactly the same as the steady solution, it holds when $\xi = 1$ that $f_m(\eta, 1) = 0$, $m = 1, 2, 3, \ldots$. Thus, when $\xi = 1$, the solution series is convergent for all $\hbar$. However, for $\xi \neq 1$ such as $\xi = 0$, we have to investigate the influence of $\hbar$ on the convergence of the solution series. To do so, we first consider $f''(0, \xi)$, which relates the local skin friction coefficient C_f^x and thus has an important physical meaning (Vajravelu and Van Gorder [59]). Regarding $\hbar$ as an unknown parameter, we can plot curves of $f''(0, \xi)$ via $\hbar$ for different ξ, called the $\hbar$ curves of $f''(0, \xi)$. For example, the $\hbar$ curve of $f''(0, \xi)$ at $\xi = 0$ is as shown in Fig. 4.11.

This $\hbar$-curve has a parallel line segment that corresponds to a region of $-0.35 < \hbar < -0.15$, denoted by R $\hbar$. The series of $f''(0, \xi)$ converges if $\hbar$ is chosen in this region.

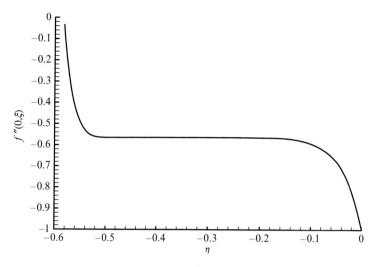

Figure 4.11 $\hbar$ curve of the 30th-order homotopy analysis method approximation to $f''(0)$ when $\xi = 0$ [59].

Table 4.4 The analytic approximations of $f''(0,0)$ by means of $\hbar = -1/4$ [59]

Order of Approximations	$f''(0,0)$
5	-0.69303
10	-0.60114
15	-0.57440
20	-0.56693
25	-0.56491
30	-0.56438
35	-0.56424
40	-0.56420
45	-0.56419
50	-0.56419

Indeed, the convergent result of $f''(0,\xi)$ is obtained when $\hbar = -1/4$ and $\xi = 0$, as shown in Table 4.4. It agrees well with the exact result $f''(0,0) = -1/\sqrt{\pi} \approx -0.56419$. Besides, the convergence can be greatly accelerated by means of the so-called homotopy–Padé method [81], as shown in Table 4.5 (Vajravelu and Van Gorder [59]).

Furthermore, it is found that when $\hbar = -1/4$ and $\xi = 0$, the 20th-order approximation and [3,3] homotopy–Padé approximation of the velocity profile $f'(\eta, 0)$ agree well with the exact solution in the whole region $0 \le \eta < +\infty$, as shown in Fig. 4.12. Similarly, given $\xi \in [0, 1]$, we can find a proper value of $\hbar$ to ensure that the solution series

Table 4.5 The $[m, n]$ homotopy–Padé approximations of $f''(0,0)$ [59]

m	$f''(0,0)$
5	−0.56415
10	−0.56418
15	−0.56419
20	−0.56419
25	−0.56419
30	−0.56419

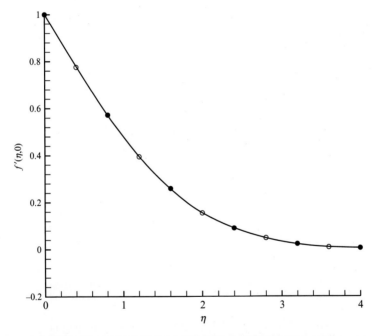

Figure 4.12 Comparison of $f'(\eta,0)$ of the exact solution (solid line) when $\xi = 0$ with a 20th-order approximation (open circle) when $\hbar = -1/4$ and the [3,3] homotopy–Padé approximation (filled circle) [59].

is convergent. It is found that when $\hbar = -1/4$, the solution series is convergent for any value of $\xi \in [0, 1]$ in the whole region $0 \leq \eta < +\infty$, as shown in Fig. 4.13 (Vajravelu and Van Gorder [59]).

    The corresponding local skin friction at the dimensionless time $\tau \in [0, +\infty]$ agrees with the numerical result, as shown in Fig. 4.14. Thus, by means of choosing $\hbar = -1/4$, we obtain an accurate analytic solution uniformly valid for all time $0 \leq \tau < +\infty$ in the whole region $0 \leq \eta < +\infty$. It should be emphasized that the solution series diverges

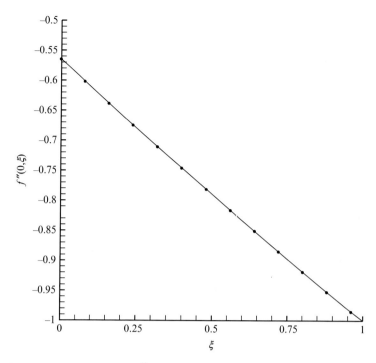

Figure 4.13 The approximation of $f''(0, \xi)$ for $0 \leq \xi \leq 1$ when $\hbar = -1/4$. Here are the 20-term (circles) and 30-term (solid line) approximate homotopy analysis method solutions [59].

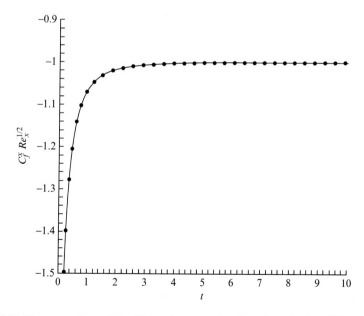

Figure 4.14 The comparison of the 30th-order approximation of $C_f^x \sqrt{Re_x}$ at different dimensionless time $\tau = at$ (30-term homotopy analysis method approximation) with the numerical result (solid line) [59].

when $\hbar = -1$ but converges when $\hbar = -1/4$. So, it is the auxiliary parameter $\hbar$ that provides us with a simple way to ensure the convergence of the solution series. This is the advantage of the homotopy analysis method (Vajravelu and Van Gorder [59]).

References

[1] Prandtl L. Über Flüssigkeitsbewegung bei sehr kleiner reibung, in: proceedings of the third international mathematical congress; 1904.

[2] Blasius H. Grenzschichten in flüssigkeiten mit kleiner reibung. Z Angew Math Phys 1908;56:1–37.

[3] Pal D, Shivakumara IS. Mixed cnvection heat transfer from a vertical heated plate embedded in a sparsely packed porous medium. Int J Appl Mech Eng 2006;11:929–39.

[4] Murthy PVSN, Mukherjee S, Krishna PVSSSR, Srinivasacharya D. Mixed convection heat and mass transfer in a doubly stratified non-Darcy porous medium. Int J Appl Mech Eng 2007;12:109–24.

[5] Soundalgekar VM, Takhar HS, Vighnesam NV. The combined free and forced convection flow past a semiinfinite plate with variable surface temperature. Nucl Eng Des 1988;110:95–8.

[6] Watanabe T, Pop I. Hall effects on magnetohydrodynamic boundary layer flow over a continuous moving flat plate. Acta Mech 1995;108:35–47.

[7] Elbashbeshy EMA, Bazid MAA. Heat transfer over a continuously moving plate embedded in non-Darcian porous medium. Int J Heat Mass Tran 2000;43(17):3087–92.

[8] Anjali Davi SP, Kandasamy R. Effects of chemical reaction, heat and mass transfer on MHD flow past a semi infinite plate. Z Angew Math Mech 2000;80:697–700.

[9] Damesh RA, Duwairi HM, Al-Odat M. Similarity analysis of magnetic field and radiation effects on forced convection flow. Turkish J Eng Env Sci 2006;30:83–9.

[10] Mukhopadhyay S, Layek GC. Radiation effect on forced convective flow and heat transfer over a porous plate in a porous medium. Meccanica 2009;44:587–97.

[11] Bhattacharyya K, Layek GC. Similarity solution of MHD boundary layer flow with diffusion and chemical reaction over a porous flat plate with suction/blowing. Meccanica 2012;47:1043–8.

[12] Mukhopadhyay S, De PR, Bhattacharyya K, Layek GC. Forced convective flow and heat transfer over a porous plate in a Darcy-Forchheimer porous medium in presence of radiation. Meccanica 2012;47:153–61.

[13] Brewster MQ. Thermal radiative transfer properties. New York: John Wiley and Sons; 1972.

[14] Damesh RA. Magnetohydrodynamics—mixed convection from radiate vertical isothermal surface embedded in a saturated porous media, Trans. ASME. J Appl Mech 2006;73:54–9.

[15] Batchelor GK. An introduction to fluid dynamics. London: Cambridge University in Press; 1987, 597.

[16] Mukhopadhyay S, Vajravelu K, Gorder RAV. Chemically reactive so lute transfer in a moving fluid over a moving surface. Acta Mech 2013;224:513–23.

[17] Cortell R. Flow and heat transfer in a moving fluid over a moving flat surface. Theor Comput Fluid Dynam 2007;21:435–46.

[18] Ishak A, Nazar R, Pop I. The effects of transpiration on the flow and heat transfer over a moving permeable surface in a parallel stream. Chem Eng J 2009;148:63–7.

[19] Mukhopadhyay S, Bhattacharyya K, Layek GC. Steady boundary layer flow and heat transfer over a porous moving plate in presence of thermal radiation. Int J Heat Mass Tran 2011;54:2751–7.

[20] Vajravelu K, Mohapatra RN. On fluid dynamic drag reduction in some boundary layer flows. Acta Mech 1990;81:59–68.

[21] Van Gorder RA. Two-dimensional Blasius viscous flow of a power-law fluid over a semi-infinite flat plane. J Math Phys 2010;51:112901.

[22] Bertolotti FP, Herbert TH, Spalart PR. Linear and nonlinear stability of the Blasius boundary layer. J Fluid Mech 1992;242:441–74.

[23] Weidman PD, Kubitschek DG, Davis AMJ. The effect of transpiration on self-similar boundary layer flow over moving surfaces. Int J Eng Sci 2006;44:730–7.

[24] Merkin JH. A note on the similarity equations arising in free convection boundary layers with blowing and suction. J Appl Math Phys (ZAMP) 1994;45:258–74.

[25] Postelnicu A, Pop I. Falkner-Skan boundary layer flow of a power-law fluid past a stretching wedge. Appl Math Comp 2011;217:4359–68.

[26] Mukhopadhyay S. Effects of slip on unsteady mixed convective flow and heat transfer past a porous stretching surface. Nucl Eng Des 2011;241:2660–5.

[27] Subhashini SV, Sumathi R, Pop I. Dual solutions in a double-diffusive MHD mixed convection flow adjacent to a vertical plate with prescribed surface temperature. Int J Heat Mass Tran 2013;56:724–31.

[28] Cao K, Baker J. Slip effects on mixed convective flow and heat transfer from a vertical plate. Int J Heat Mass Tran 2009;52:3829–41.

[29] Bhattacharyya K, Mukhopadhyay S, Layek GC. Similarity solution of mixed convective boundary layer slip flow over a vertical plate. Ain Shams Eng J 2013;4(2): 299–305.

[30] Patil PM, Anil Kumar D, Roy S. Unsteady thermal radiation mixed convection flow from a moving vertical plate in a parallel free stream: effect of Newtonian heating. Int J Heat Mass Tran 2013;62:534–40.

[31] Mukhopadhyay S, Mandal IC, Hayat T. Mixed convection slip flow with heat transfer and porous medium. J Porous Med 2014;17(11):1007–17.

[32] Bhattacharyya K, Mukhopadhyay S, Layek GC. Steady boundary layer slip flow and heat transfer over a flat porous plate embedded in a porous media. J Petrol Sci Eng 2011;78:304–9.

[33] Gupta AS. Laminar free convection flow of an electrically conducting fluid from a vertical plate with uniform surface heat flux and variable wall temperature in presence of magnetic field. J Appl Math Phys (ZAMP) 1963;13:324–32.

[34] Emery AF. The effect of magnetic field upon the free convection of a conducting fluid. J Heat Transfer Ser C 1963;85:119–24.

[35] Takhar HS. Hydromagnetic free convection from a flat plate. Indian J of Phys 1971; 45:289–311.

[36] Cheng P, Minkowycz WJ. Free convection about a vertical flat plate embedded in a porous medium with application to heat transfer from a disk. J Geophys Res 1963;82: 2040–4.

[37] Cheng P. The influence of lateral mass flux on a free convection boundary layers in saturated porous medium. Int J Heat Mass Tran 1977;20:201–6.

[38] Wilks G. Combined forced and free convection flow on vertical surfaces. Int J Heat Mass Tran 1973;16:1958–64.

[39] Boutros YZ, Abd-el-Malek MB, Badran NA. Group theoretic approach for solving time independent free-convective boundary layer flow on a nonisothermal vertical flat plate. Arch Mech 1990;42:377–95.

[40] Chen TS, Strobel FA. Buoyancy effects in boundary layer adjacent to a continuous moving horizontal flat plate. J Heat Transfer 1980;102:170–2.

[41] Ramachandran BF, Armaly BF, Chen TS. Correlation for laminar mixed convection on boundary layers adjacent to inclined continuous moving sheets. Int J Heat Mass Tran 1987;30:2196–9.

[42] Lee SL, Tsai JS. Cooling of a continuous moving sheet of finite thickness in the presence of natural convection. Int J Heat Mass Tran 1990;33:457–64.

[43] Mukhopadhyay S, Layek GC, Gorla RSR. Radiation effects on MHD free-convective flow past a semi infinite vertical plate with a power-law velocity distribution. Int J Fluid Mech Res 2010;37(6):567–81.

[44] Chiam TC. Hydromagnetic flow over a surface stretching with a power-law velocity. Int J Eng Sci 1993;33:429–35.

[45] Mukhopadhyay S, Layek GC, Gorla RSR. MHD combined convective flow past a stretching surface. Int J Fluid Mech Res 2007;34:244–57.

[46] Magyari E, Keller B. Exact solutions for self-similar boundary-layer flows induced by permeable stretching walls. Eur J Mech B Fluids 2000;19:109–22.

[47] Siekman J. The laminar boundary layer along a flat plate. Z Flugwiss 1962;10:278–81.

[48] Klemp JB, Acrivos A. The moving-wall boundary layer with reverse flow. J Fluid Mech 1976;76:363–81.

[49] Abdulhafez TA. Skin friction and heat transfer on a continuous flat surface moving in a parallel free stream. Int J Heat Mass Tran 1985;28:1234–7.

[50] Chappidi PR, Gunnerson FS. Analysis of heat and momentum transport along a moving surface. Int J Heat Mass Tran 1989;32:1383–6.

[51] Hussaini MY, Lakin WD, Nachman A. On similarity solutions of a boundary-layer problem with an upstream moving wall. SIAM J Appl Math 1987;47:699–709.

[52] Lin HT, Haung SF. Flow and heat transfer of plane surface moving in parallel and reversely to the free stream. Int J Heat Mass Tran 1994;37:333–6.

[53] Sparrow EM, Abraham JP. Universal solutions for the streamwise variation of the temperature of a moving sheet in the presence of a moving fluid. Int J Heat Mass Tran 2005;48:3047–56.

[54] Afzal N, Badaruddin A, Elgarvi AA. Momentum and heat transport on a continuous flat surface moving in a parallel stream. Int J Heat Mass Tran 1993;36:3399–403.

[55] Mukhopadhyay S. Heat transfer in a moving fluid over a moving non-isothermal flat surface. Chin Phys Lett 2011;28(12):124706.

[56] Mondal IC, Mukhopadhyay S, Gorla RSR. Dual solutions for the boundary layer flow of a nanofluid over a moving surface. Int J Micro-Nano Scale Transport 2011;2 (4):221–33.

[57] Lakshmi Narayana PA, Murthy PVSN, Gorla RSR. Soret-driven thermosolutal convection induced by inclined thermal and solutal gradients in a shallow horizontal layer of a porous medium. J Fluid Mech 2008;612:1–19.

[58] Liao S. An analytic solution of unsteady boundary-layer flows caused by an impulsively stretching plate. Commun Nonlinear Sci Numer Simul 2006;11:326–9.

[59] Vajravelu K, Van Gorder RA. Nonlinear flow phenomena and homotopy analysis: fluid flow and heat transfer. Beijing: Higher Education Press; 2012, and Springer-Verlag, Berlin Heidelberg.

[60] Sakiadis BC. Boundary layer behaviour on continuous solid surface. I: the boundary layer equation for two dimensional and asymmetric flow. AIChE J 1961;7:26–8.

[61] Crane LJ. Flow past a stretching plate. Z Angew Math Phys 1970;21:645–7.

[62] Banks WHH. Similarity solutions of the boundary-layer equations for a stretching wall. J Mech Theor Appl 1983;2:375–92.

[63] Grubka LJ, Bobba KM. Heat transfer characteristics of a continuous stretching surface with variable temperature. ASME J Heat Transfer 1985;107:248–50.

[64] Ali ME. Heat transfer characteristics of a continuous stretching surface. Wärme Stoffü; bertrag 1994;29:227–34.

[65] L. E. Erickson, L. T. Fan, and V. G. Fox, Heat and mass transfer on a moving continuous flat plate with suction or injection, Ind. Eng. Chem. 5 (199) 19-25.

[66] Gupta PS, Gupta AS. Heat and mass transfer on a stretching sheet with suction or blowing. Can J Chem Eng 1977;55:744–6.

[67] Chen CK, Char MI. Heat transfer of a continuous stretching surface with suction or blowing. J Math Anal Appl 1988;135:568–80.

[68] Chaudhary MA, Merkin JH, Pop I. Similarity solutions in the free convection boundary-layer flows adjacent to vertical permeable surfaces in porous media. Eur J Mech B Fluids 1995;14:217.

[69] Elbashbeshy EMA. Heat transfer over a stretching surface with variable surface heat flux. J Phys D Appl Phys 1998;31:1951–6.

[70] Magyari E, Keller B. Exact solutions for self-similar boundary-layer flows induced by permeable stretching walls. Eur J Mech B Fluids 2000;19:109.

[71] Stewartson K. On the impulsive motion of a flat plate in a viscous fluid (Part I). Q J Mech 1951;4:182.

[72] Stewartson K. On the impulsive motion of a flat plate in a viscous fluid (Part II). Q J Mech Appl Math 1973;22:143.

[73] Hall MG. The boundary layer over an impulsively started flat plate. Proc R Soc Lond A 1969;310:401.

[74] Dennis SCR. The motion of a viscous fluid past an impulsively started semi-infinite flat plate. J Inst Math Appl 1972;10:105.

[75] Watkins CB. Heat transfer in the boundary layer over an impulsively started flat plate. J Heat Transfer 1975;97:282.

[76] Seshadri R, Sreeshylan N, Nath G. Unsteady mixed convection flow in the stagnation region of a heated vertical plate due to impulsive motion. Int J Heat Mass Tran 2002;45:1345.

[77] Pop I, Na TY. Unsteady flow past a stretching sheet. Mech Resc Commun 1996;23:413–22.

[78] Wang CY, Du G, Miklavcic M, Chang CC. Impulsive stretching surface in a viscous fluid. SIAM J Appl Math 1997;57:1.

[79] Nazar N, Amin N, Pop I. Unsteady boundary layer flow due to a stretching surface into a rotating fluid. Mech Res Commun 2004;31:121.

[80] Williams JC, Rhyne TH. Boundary layer development on a wedge impulsively set in motion. SIAM J Appl Math 1980;38:215.

[81] Liao SJ. Beyond perturbation: introduction to the homotopy analysis method. Boca Raton, FL: chapman & Hall\CRC press; 2003.

[82] Liao SJ. On the homotopy analysis method for nonlinear problems. Appl Math Comput 2004;147:499–513.

[83] Liao SJ. On the analytic solution of magnetohydrodynamic flows of non-Newtonian fluids over a stretching sheet. J Fluid Mech 2003;488:189–212.

[84] Hayat T, Khan M, Ayub M. On the explicit analytic solutions of an oldroyd 6-constant fluid. Int J Eng Sci 2004;42:123–35.

[85] Hayat T, Khan M, Asghar S. Magnetohydrodynamic flow of an Oldroyd 6-constant fluid. Appl Math Comput 2004;155:417–25.

[86] Hayat T, Khan M, Asghar S. Homotopy analysis of MHD flows of an oldroyd 8-constant fluid. Acta Mech 2004;168:213–32.

[87] Liao SJ, Campo A. Analytic solutions of the temperature distribution in blasius viscous flow problems. J Fluid Mech 2002;453:411–25.

[88] Liao Sj. A uniformly valid analytic solution of 2D viscous flow past a semi-infinite flat plate. J Fluid Mech 1999;385:101–28.

[89] Ayub M, Rasheed A, Hayat T. Exact flow of a third grade fluid past a porous plate using homotopy analysis. Int J Eng Sci 2003;41:2091–103.

[90] Ifidon EO. An application of homotopy analysis to the viscous flow past a circular cylinder. J Appl Math 2009;2009:17.

[91] Allan FM, Syam MI. On the analytic solutions of the non-homogenous blasius problem. J Comput Appl Math 2005;182:362.

Part II

Further Applications

Flow past a cylinder

5

During the last few decades, flow, heat, and mass transfer at different shapes of objects drew significant interest (Pal and Mondal [1]). Among them, flow past a stretching surface was widely studied. The study of hydrodynamic flow over cylinders has gained considerable attention because of its applications to industries and important bearings on several technological processes. Because of a wide range of applications in coating of wires and polymer fiber spinning, the concept of heat convection in the cylinders have been a field of interest for many theoretical and experimental researchers. The flow over cylinders is considered to be two-dimensional if the radius is large compared to the boundary-layer thickness. On the other hand, for a thin or slender cylinder, the radius of the cylinder may be of the same order as the boundary-layer thickness. Therefore, the flow may be considered as axisymmetric instead of two-dimensional (Datta et al. [2], Kumari and Nath [3]).

5.1   Flow past a stretching cylinder

Most of the investigators restricted their analyses to two-dimensional flow over a stretching sheet, but not much has been done for the more intricate problem of the axisymmetric flow due to a stretching cylinder. With an overview to the aforementioned industrial processes related to stretching sheet problems, it can be concluded that many of them such as hot rolling, wire coating, food processing, etc. involve cylindrical geometries.

The study of steady flow in a viscous and incompressible fluid outside a stretching hollow cylinder in an ambient fluid at rest was carried out by Wang [4]. The effect of slot suction/injection over a thin cylinder as studied by Datta et al. [2] and Kumari and Nath [3] may be useful in the cooling of nuclear reactors during emergency shutdown, where a part of the surface can be cooled by injecting a coolant (Ishak et al. [5]). Lin and Shih [6,7] considered the laminar boundary layer and heat transfer along horizontally and vertically moving cylinders with constant velocity and found that the similarity solutions could not be obtained because of the curvature effect of the cylinder. Ishak and Nazar [8] showed that the similarity solutions may be obtained by assuming that the cylinder is stretched with linear velocity in the axial direction and claimed that their study may be regarded as the extension of the papers by Grubka and Bobba [9] and Ali [10] from a stretching sheet to a stretching cylinder. Ishak et al. [11] examined the effects of a magnetic field on steady flow and heat transfer of a fluid along a stretching cylinder without considering the transverse curvature. But the impact of transverse curvature is vital in many applications such as wire and fibre drawing wherein accurate prediction is expected and a thick boundary layer can exist on slender or near slender bodies. Afterwards, Bachok and Ishak [12] presented a numerical study on the flow and heat transfer over a horizontal cylinder considering the effect of

Fluid Flow, Heat and Mass Transfer at Bodies of Different Shapes. http://dx.doi.org/10.1016/B978-0-12-803733-1.00005-3

the transverse curvature parameter and prescribed surface heat flux. Vajravelu et al. [13] explored the influence of the temperature-dependent thermal conductivity on flow and thermal boundary layers along a stretching slender cylinder. They also considered the external heat source and transverse magnetic field for two various thermal boundary conditions. Here we shall present the results of Mukhopadhyay [14], who considered the axisymmetric flow and heat transfer of viscous fluid along a stretching cylinder in the presence of a magnetic field.

5.1.1 *Mathematical analysis of the problem*

The transverse curvature parameter is defined as the ratio of the boundary-layer thickness and cylinder radius. For a very small transverse curvature parameter, the problem can be solved in two dimensions, but for a thin cylinder, the radius order can be the same as the order of the boundary-layer thickness. As a result, the problem can be assumed axisymmetric. In the latter case, the surface transverse curvature parameter appears in the governing equations.

Let us consider an electrically conducting Newtonian fluid around a slender cylinder with a prescribed surface temperature as shown in Fig. 5.1. Stretching the surface causes an incompressible laminar boundary-layer flow towards x-direction that is also affected by an external radial magnetic field.

Consider the steady axisymmetric flow of an incompressible viscous fluid along a stretching cylinder in the presence of a uniform magnetic field (see Fig. 5.1). The x-axis is measured along the axis of the tube and the r-axis is measured in the radial direction. It is assumed that the uniform magnetic field of intensity B_0 acts in the radial direction. The magnetic Reynolds number is assumed to be small so that the induced magnetic field is negligible in comparison with the applied magnetic field. The continuity, momentum, and energy equations governing such flows are written as

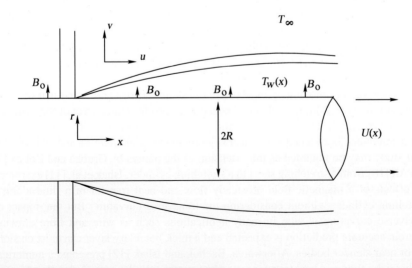

Figure 5.1 Sketch of the horizontal cylinder [14].

$$\frac{\partial(ru)}{\partial x} + \frac{\partial(rv)}{\partial r} = 0, \tag{5.1}$$

$$u\frac{\partial u}{\partial x} + v\frac{\partial u}{\partial r} = \frac{v}{r}\frac{\partial}{\partial r}\left(r\frac{\partial u}{\partial r}\right) - \frac{\sigma B_0^2}{\rho}u, \tag{5.2}$$

$$u\frac{\partial T}{\partial x} + v\frac{\partial T}{\partial r} = \frac{\kappa}{r}\frac{\partial}{\partial r}\left(r\frac{\partial T}{\partial r}\right), \tag{5.3}$$

where u and v are the components of velocity respectively in the x and r directions, $v = \frac{\mu}{\rho}$ is the kinematic viscosity, ρ is the fluid density, μ is the coefficient of fluid viscosity, σ is the electrical conductivity of the medium, B_0 is the uniform magnetic field, κ is the thermal diffusivity of the fluid, T is the fluid temperature.

The appropriate boundary conditions for the problem are given by

$$u = U(x) + B_1 v\frac{\partial u}{\partial r}, v = 0, T = T_w(x) \text{ at } r = R, \tag{5.4a}$$

$$u \to 0, T \to T_\infty \text{ as } r \to \infty. \tag{5.4b}$$

Here $U(x) = U_0\frac{x}{L}$ is the stretching velocity, $T_w(x) = T_\infty + T_0\left(\frac{x}{L}\right)^N$ is the prescribed surface temperature, U_0 and T_0 are the reference velocity and temperature respectively, T_∞ is the ambient temperature, L is the characteristic length, N is the temperature exponent, and B_1 is the velocity slip.

5.1.2 Solution procedure

The continuity equation is automatically satisfied by the introduction of the stream function ψ as

$$u = \frac{1}{r}\frac{\partial \psi}{\partial r}, v = -\frac{1}{r}\frac{\partial \psi}{\partial x}. \tag{5.5}$$

Introducing the similarity variable and similarity transformations as

$$\eta = \frac{r^2 - R^2}{2R}\left(\frac{U}{vx}\right)^{\frac{1}{2}}, \psi = (Uvx)^{\frac{1}{2}}Rf(\eta), \theta(\eta) = \frac{T - T_\infty}{T_w - T_\infty}, \tag{5.6}$$

and upon substitution of (5.6) in equations (5.2), (5.3), (5.4a), and (5.4b), the governing equations and the boundary conditions reduce to

$$(1 + 2M\eta)f''' + 2Mf'' + ff'' - f'^2 - D^2f' = 0, \tag{5.7}$$

$$(1 + 2M\eta)\theta'' + 2M\theta' + Pr\left(f\theta' - Nf'\theta\right) = 0, \tag{5.8}$$

$$f'(\eta) = 1 + Bf''(\eta), f(\eta) = 0, \theta = 1 \text{ at } \eta = 0, \tag{5.9a}$$

and

$$f'(\eta) \to 0, \theta(\eta) \to 0 \quad \text{as} \quad \eta \to \infty, \tag{5.9b}$$

where the prime denotes differentiation with respect to η, $B = B_1 \sqrt{\frac{U_0 \nu}{L}}$ is the slip parameter, $D^2 = \frac{\sigma B_0^2 L}{\rho U_0}$, D is the magnetic parameter, and $M = \left(\frac{\nu L}{U_0 R^2}\right)^{1/2}$ is the curvature parameter. The no-slip case is recovered for $B = 0$.

One can note that if $M = 0$ (i.e., $R \to \infty$), the problem under consideration (with $B = 0$, $D = 0$) reduces to the boundary-layer flow along a stretching flat plate considered by Ali [10], with $m = 1$ in that paper. Moreover, when $M = 0$ (stretching flat plate) subject to (5.9) with $D = 0$ (in the absence of a magnetic field) and $B = 0$ (for no-slip case), then the analytical solutions of equations (5.7) and (5.8) are given by Crane [15] and Grubka and Bobba [9], respectively.

The closed-form solution of equation (5.7) with $M = 0$, $B = 0$, and $D = 0$ is $f(\eta) = 1 - e^{-\eta}$.

The analytical solution of equation (5.7) for $M = 0$ (for stretching flat plate) in the presence of a magnetic field with the no-slip boundary condition (i.e., for $B = 0$) is given by

$$f(\eta) = \frac{1}{\sqrt{D^2 + 1}} \left(1 - e^{-\sqrt{D^2 + 1}\eta}\right). \tag{5.10}$$

It is easy to find that $f'(\eta) = e^{-\sqrt{D^2 + 1}\eta}$ and the value of $f''(0) = -\sqrt{D^2 + 1}$.

Now let us find the analytic solution for equation (5.7) with $M = 0$ (for a flat plate) in the presence of a magnetic field and with slip at the boundary, that is, the solution of the equation

$$f''' + ff'' - f'^2 - D^2 f' = 0 \tag{5.11}$$

subject to the boundary conditions

$$f(0) = 0, f'(0) = 1 + Bf''(0) \quad \text{and} \quad f'(\infty) \to 0. \tag{5.12}$$

Let us assume a solution of the form $f(\eta) = a + be^{-\alpha\eta}$.
Substituting this in equation (5.11) we get

$$a = \frac{1}{\alpha + B\alpha^2}, b = -\frac{1}{\alpha + B\alpha^2},$$

where α is the root of the equation

$$B\alpha^3 + \alpha^2 - BD^2\alpha - 1 - D^2 = 0. \tag{5.13}$$

Thus, the solution takes the following form:

$$f(\eta) = \frac{1}{\alpha + B\alpha^2} - \frac{1}{\alpha + B\alpha^2} e^{-\alpha\eta}.$$

Following Fang et al. [16], equation (5.13) can be converted to an equation $\beta^3 + r\beta + s = 0$, where

$$\beta = \alpha + \frac{1}{3B}, r = -\frac{1}{3B^2} - D^2, s = \frac{2}{27B^3} + \frac{D^2}{3B} - \frac{1+D^2}{B}.$$

Now the roots of the above equation are given by

$$\beta_1 = E + F, \beta_{2,3} = -\frac{1}{2}(E+F) \pm i\frac{\sqrt{3}}{2}(E-F),$$

where

$$E = \sqrt[3]{-\frac{s}{2} + \sqrt{G}}, F = \sqrt[3]{-\frac{s}{2} - \sqrt{G}}, i^2 = -1, G = \left(\frac{r}{3}\right)^2 + \left(\frac{s}{2}\right)^2.$$

The positive real roots of α are physically relevant. For $D = 0$, this solution reduces to the solutions obtained by Wang [17] and Andersson [18].

The solution of equation (5.12) for different values of the slip parameter (B) and magnetic parameter (D) are presented in Table 5.1, which agrees with that of Fang et al. [16].

For a stretching flat plate in the no-slip case and for $N = 2$, the energy equation becomes

$$\theta^{//} + Prf\theta^{/} - 2Prf^{/}\theta = 0. \tag{5.14}$$

For the solution of equation (5.14) subject to the boundary conditions

$$\theta(0) = 1, \theta(\infty) \to 0, \tag{5.15}$$

following Anjali Devi and Ganga [19], let us introduce an independent variable ξ given by

$$\xi = \frac{-Pr}{\alpha^2}e^{-\alpha\eta}. \tag{5.16}$$

Table 5.1 **Solution for α for different values of B and D**

	Fang et al. [16]		Present study	
B	$D = 0.5$	$D = 2$	$D = 0.5$	$D = 2$
0	1.1180	2.2361	1.1180	2.2361
0.5	0.9619	2.1179	0.9619	2.1179

Reprinted from Ref. [14].

Substituting (5.10) and (5.16) in equation (5.14), we get

$$\xi\frac{d^2\theta}{d\xi^2} + [(1-K) - \xi]\frac{d\theta}{d\xi} + 2\theta = 0. \tag{5.17}$$

The corresponding boundary conditions become $\theta(\xi=0) = 1, \theta\left(\xi=\frac{-Pr}{\alpha^2}\right) = 1,$
where $K = \frac{Pr}{\alpha^2}[\alpha^2 - D^2]$. The solution of equation (5.17) can be written in terms of
the confluent hypergeometric function as

$$\theta(\xi) = \frac{\xi^K {}_1F_1(-2+K;1+K;\xi)}{\left(\frac{-Pr}{\alpha^2}\right)^K {}_1F_1\left(-2+K;1+K;\frac{-Pr}{\alpha^2}\right)}.$$

In terms of η, θ can be expressed as

$$\theta(\eta) = \frac{e^{-\alpha K\eta} {}_1F_1\left(-2+K;1+K;\frac{-Pr}{\alpha^2}e^{-\alpha\eta}\right)}{{}_1F_1\left(-2+K;1+K;\frac{-Pr}{\alpha^2}\right)}. \tag{5.18}$$

The heat transfer rate at the surface is given by

$$\theta'(0) = -\alpha K + \frac{Pr}{\alpha}\left(\frac{K-2}{K+1}\right)\frac{{}_1F_1\left(-1+K;2+K;\frac{-Pr}{\alpha^2}\right)}{{}_1F_1\left(-2+K;1+K;\frac{-Pr}{\alpha^2}\right)}. \tag{5.19}$$

5.1.3 Numerical results and discussion

To analyze the results, numerical computation has been carried out, and the numerical and exact solutions for a stretching flat plate in the no-slip case are presented in Table 5.2.

In Fig. 5.2(a), horizontal velocity profiles are shown for different values of M. The horizontal velocity curves show that the rate of transport decreases with the increasing distance (η) of the sheet. In all cases, the velocity vanishes at some large distance from the sheet (at $\eta = 10$). The velocity finally increases with increasing values of M. The velocity gradient at the surface is larger for larger values of M, which produces a larger skin friction coefficient. As the lateral surface of the cylinder changes according to the curvature parameter $M = 0, 0.25, 0.5, 0.75,$ and 1, the curvature parameter has substantial effects on velocity and temperature distributions.

Effects of the curvature parameter on the temperature distribution are presented in Fig. 5.2(b). Temperature is found to decrease with the increasing curvature parameter M. The temperature gradient at the surface increases as M increases, and the thermal

Table 5.2 **Values of $[f''(0)]$ obtained from analytical and numerical solutions**

D	Analytical solution	Numerical solution	Error
0	−1.0000000	−0.99005806	−0.00994194
0.5	−1.1180340	−1.1056039	−0.012430088
1	−1.4142135	−1.3943545	−0.019859062
1.5	−1.802775638	−1.7705669	−0.032208737

Reprinted from Ref. [14].

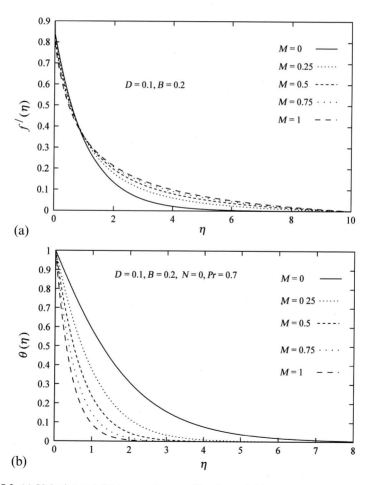

(a)

(b)

Figure 5.2 (a) Velocity and (b) temperature profiles for variable curvature parameter M. Reprinted from Ref. [14].

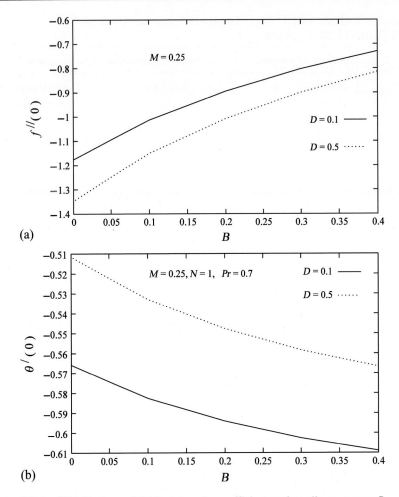

Figure 5.3 (a) Skin-friction and (b) heat transfer coefficient against slip parameter B. Reprinted from Ref. [14].

boundary-layer thickness decreases with increasing M. Thus, the surface heat transfer rate increases as M increases. Hence, the local Nusselt number $\left[Nu = \dfrac{-\theta'(0)}{\sqrt{Re}}\right]$ increases with increasing M.

Figure 5.3(a) displays the behavior of the skin friction coefficient with slip parameter B for two values of the magnetic parameter D. Skin friction increases with increasing slip parameter, but it decreases with an increasing magnetic parameter. When slip occurs, the flow velocity near the stretching wall is no longer equal to the stretching velocity of the wall. With the increase in B, the slip velocity increases and consequently fluid velocity decreases because under the slip condition, the pulling of the stretching wall can only be partly transmitted to the fluid. From the figure, it is very clear that shear stress at the wall is negative. Physically, a negative sign for $f''(0)$ implies that surface exerts a dragging force on the fluid and a positive sign implies the opposite. This is consistent with the present case, as a stretching cylinder that induces the flow is considered here. Figure 5.3(b) displays the nature of heat transfer coefficient with slip parameter B. It is found that the heat transfer coefficient increases with increasing magnetic parameter D but decreases with slip parameter B. It is obvious that the presence of a magnetic field causes higher restriction to the fluid, which reduces the fluid velocity. Actually, the magnetic field opposes the transport phenomena. This is due to the fact that with increasing D, the Lorentz force associated with the magnetic field increases and it produces more resistance to the transport phenomena. But the magnetic field enhances the temperature at all points, leading to increase in thermal boundary-layer thickness.

5.2 Flow past a vertical cylinder

Flow and heat transfer due to a stretching cylinder in a quiescent or moving fluid is important in a number of industrial manufacturing processes that include both metal and polymer sheets. It is worth mentioning that there are several practical applications in which significant temperature differences between the body surface and the ambient fluid exist. The temperature differences cause density gradients in the fluid medium, and free convection effects become more important in the presence of gravitational force. Mixed convection flows or coupled forced and natural convection flows arise in many transport processes both in natural and engineering applications (Mukhopadhyay [20]). Such processes occur when the effects of buoyancy forces in forced convection or the effects of forced flow in natural convection become much more significant. There arise some situations where the stretching cylinder moves vertically in the cooling liquid. In this situation, the fluid flow and the heat transfer characteristic are determined by two mechanisms, namely, the motion of the stretching cylinder and the buoyancy force. The thermal buoyancy generated due to heating/cooling of a vertically moving stretching cylinder has a large impact on the flow and heat transfer characteristics (see, [20]). Convection heat transfer and fluid flow through a porous medium is a phenomenon of great interest from both a theoretical and practical point of view because of its applications in many engineering and geophysical fields such as geothermal and petroleum resources, solid matrix heat exchanges, thermal insulation drying of porous solids, enhanced oil recovery, cooling of nuclear reactors, and other practical interesting designs (Mukhopadhyay and Layek [21], Mukhopadhyay et al. [22]).

The study of hydrodynamic flow and heat transfer in porous media becomes much more interesting because of its vast applications on the boundary-layer flow control. Heat removal from nuclear fuel debris, underground disposal of radioactive waste material, storage of food stuffs, and exothermic and/or endothermic chemical reactions and dissociating fluids in the packed-bed reactors, etc. are some porous media applications. It is well known that Darcy's law is an empirical formula relating the pressure gradient, the bulk viscous fluid resistance, and the gravitational force for a forced convective flow in a porous medium. Deviations from Darcy's law occur when the Reynolds number based on the pore diameter is within the range of 1–10 (Ishak et al. [23]).

Now we shall present the results of Mukhopadhyay [20].

5.2.1 Mathematical formulation of the problem and solution procedure

Consider the steady axisymmetric mixed convection flow of an incompressible viscous fluid along a vertical stretching cylinder in a porous medium (see Fig. 5.4). The continuity, momentum, and energy equations governing such flow are written as

$$\frac{\partial(ru)}{\partial x}+\frac{\partial(rv)}{\partial r}=0, \tag{5.20}$$

$$u\frac{\partial u}{\partial x}+v\frac{\partial u}{\partial r}=\frac{v}{r}\frac{\partial}{\partial r}\left(r\frac{\partial u}{\partial r}\right)-\frac{v}{k}u+g\beta(T-T_\infty), \tag{5.21}$$

$$u\frac{\partial T}{\partial x}+v\frac{\partial T}{\partial r}=\frac{\kappa}{r}\frac{\partial}{\partial r}\left(r\frac{\partial T}{\partial r}\right), \tag{5.22}$$

where u and v are the components of velocity respectively in the x and r directions, $v=\frac{\mu}{\rho}$ is the kinematic viscosity, ρ is the fluid density, μ is the coefficient of fluid viscosity, k is the permeability of the medium, κ is the thermal diffusivity of the fluid, T is the fluid temperature, β is the volumetric coefficient of thermal expansion, g is the gravity field, and T_∞ is the ambient temperature.

The appropriate boundary conditions for the problem are given by

$$u=U(x),v=0,T=T_w(x) \quad \text{at} \quad r=R, \tag{5.23a}$$

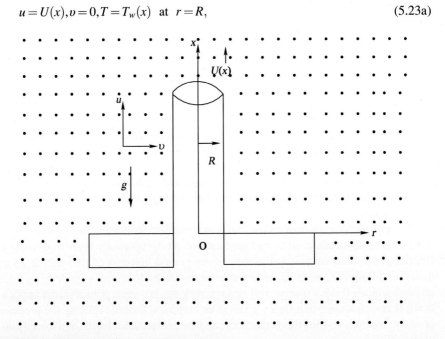

Figure 5.4 Sketch of the physical flow problem.
Reprinted from Ref. [20].

$$u \rightarrow 0, T \rightarrow T_\infty \quad \text{as} \quad r \rightarrow \infty. \tag{5.23b}$$

Here $U(x) = U_0 \frac{x}{L}$ is the stretching velocity, $T_w(x) = T_\infty + T_0 \left(\frac{x}{L}\right)^N$ is the prescribed surface temperature (for forced convection case), N is the temperature exponent, and $N = 1$ is considered for the mixed convection case. U_0 and T_0 are the reference velocity and temperature respectively and L is the characteristic length.

The continuity equation is automatically satisfied by the introduction of the stream function ψ as $u = \frac{1}{r} \frac{\partial \psi}{\partial r}, v = -\frac{1}{r} \frac{\partial \psi}{\partial x}$.

We introduce the similarity variables

$$\eta = \frac{r^2 - R^2}{2R} \left(\frac{U}{\nu x}\right)^{\frac{1}{2}}, \psi = (U\nu x)^{\frac{1}{2}} Rf(\eta), \theta(\eta) = \frac{T - T_\infty}{T_w - T_\infty}, \tag{5.24}$$

and upon substitution of (5.24) in equations (5.21), (5.22), (5.23a), and (5.23b), the governing equations and the boundary conditions reduce to

$$(1 + 2M\eta)f''' + 2Mf'' + ff'' - f'^2 - k_1 f' + \lambda \theta = 0, \tag{5.25}$$

$$(1 + 2M\eta)\theta'' + 2M\theta' + Pr\left(f\theta' - f'\theta\right) = 0, \tag{5.26}$$

$$f' = 1, f = 0, \theta = 1 \quad \text{at} \quad \eta = 0, \tag{5.27a}$$

and

$$f' \rightarrow 0, \theta \rightarrow 0 \quad \text{as} \quad \eta \rightarrow \infty, \tag{5.27b}$$

where the prime denotes differentiation with respect to η, $k_1 = \frac{\nu L}{U_0 k}$ is the permeability parameter of the porous medium, $M = \left(\frac{\nu L}{U_0 R^2}\right)^{\frac{1}{2}}$ is the curvature parameter, and $\lambda = \frac{g\beta T_0 L}{U_0^2}$ is the mixed convection parameter. The case of a non-porous medium is recovered for $k_1 = 0$. Here, k_1^{-1} will reflect the effect of Darcian flow on the present problem.

5.2.2 Analysis of results and concluding remarks

Figures 5.5(a) and (b) displays the effects of the mixed convection parameter on velocity, shear stress, and temperature for a vertical stretching cylinder. With increasing λ, the horizontal velocity is found to increase for buoyancy-aided flow ($\lambda > 0$) but decreases for buoyancy-opposed flow ($\lambda < 0$) (Fig. 5.5(a) and (b)). It is noted that λ has a substantial effect on the solutions. Also, with increasing values of the mixed convection parameter λ, the shear stress $f''(\eta)$ initially increases for buoyancy-opposed flow but decreases in the case of buoyancy-aided flow. But after a certain distance, the opposite nature of shear stress is noted (Fig. 5.5(a)). Here, $\lambda = 0$ corresponds to the

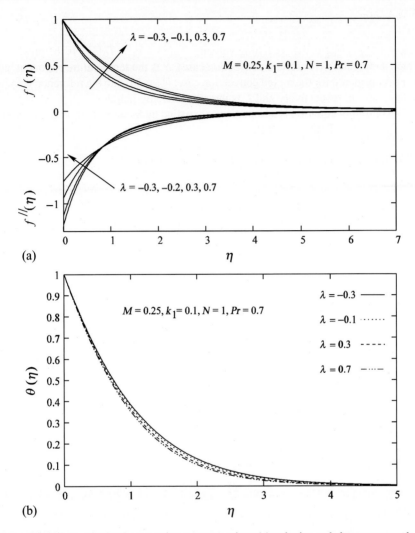

Figure 5.5 Effects of mixed convection parameter λ on (a) velocity and shear stress and (b) temperature profiles for vertical stretching cylinder.
Reprinted from Ref. [20].

forced convection case. For $\lambda > 0$, there is a favorable pressure gradient due to the buoyancy forces, which results in the flow being accelerated. Physically, $\lambda > 0$ means heating of the fluid or cooling of the surface (assisting flow), $\lambda < 0$ means cooling of the fluid or heating of the surface (opposing flow). Also, an increase in the value of λ can lead to an increase in the temperature difference $T_w - T_\infty$.

This leads to an enhancement of the velocity due to the enhanced convection currents and thus an increase in the boundary-layer thickness. Temperature decreases with increasing λ for buoyancy-aided flow but increases in the case of buoyancy-opposed flow. An increase in the value of the mixed convection parameter λ results in a decrease in the thermal boundary-layer thickness, and this results in an increase in the magnitude of the wall temperature gradient. This in turn produces an increase in the surface heat transfer rate.

Figure 5.6(a) and (b) presents the behavior of skin friction and heat transfer coefficients versus the permeability parameter k_1 of the porous medium for three values of the curvature parameter. The magnitude of the skin friction coefficient increases with increasing permeability parameter k_1 (Fig. 5.6(a)) and also with the curvature parameter M, which also supports our earlier findings. Shear stress at the wall is negative

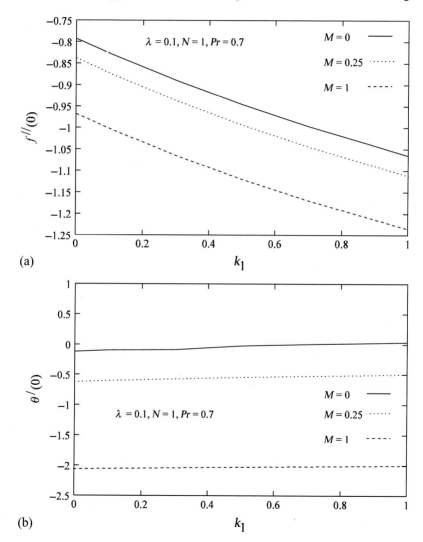

(a)

(b)

Figure 5.6 Variations of (a) skin-friction and (b) heat transfer coefficient with permeability parameter.
Reprinted from Ref. [20].

here. Physically, a negative sign for $f''(0)$ implies that the surface exerts a dragging force on the fluid and a positive sign implies the opposite. This is consistent with the present case of a stretching cylinder. The magnitude of the heat transfer coefficient decreases with increasing permeability parameter k_1. But the magnitude of the heat transfer coefficient increases with the increasing curvature parameter M (Fig. 5.6(b)).

5.3   Nanofluid boundary layer over a stretching cylinder

Nanofluids are solid–liquid composite materials consisting of solid nanoparticles or nanofibers with sizes typically of 1–100 nm suspended in liquid. These fluids represent an innovative way of increasing the thermal conductivity and, therefore, heat transfer [24]. Actually, the thermal conductivity of these fluids plays an important role on the heat transfer coefficient between the heat transfer medium and the heat transfer surface [25]. Also, it shows significant change in viscosity and specific heat compared to the base fluid. For this reason, it has attracted the attention of several researchers. The enhanced thermal behavior of nanofluids can provide a basis for an enormous innovation for heat transfer intensification, which is very important in a number of industrial sectors, including transportation, power generation, micro-manufacturing, thermal therapy for cancer treatment, chemical and metallurgical sectors, as well as heating, cooling, ventilation, and air-conditioning [26]. Nanofluids have attracted great interest recently because of reports of greatly enhanced thermal properties. For example, a small amount (<1% volume fraction) of Cu nanoparticles or carbon nanotubes dispersed in ethylene glycol or oil is reported to increase the inherently poor thermal conductivity of the liquid by 40% and 150%, respectively [27,28]. Conventional particle-liquid suspensions require high concentrations (>10%) of particles to achieve such enhancement. However, problems of rheology and stability are amplified at high concentration, precluding the widespread use of conventional slurries as heat transfer fluids. In some cases, the observed enhancement in thermal conductivity of nanofluids is orders of magnitude larger than predicted by well-established theories. Other perplexing results in this rapidly evolving field include a surprisingly strong temperature dependence of the thermal conductivity and a threefold higher critical heat flux compared with the base fluids [29,30]. These enhanced thermal properties are not merely of academic interest. If confirmed and found consistent, they would make nanofluids promising for application in thermal management. Furthermore, suspensions of metal nanoparticles are also being developed for other purposes, such as medical applications including cancer therapy. The interdisciplinary nature of nanofluid research presents a great opportunity for exploration and discovery at the frontiers of nanotechnology. We shall highlight some important results of Mukhopadhyay et al. [31].

5.3.1   Formulation of the problem

Consider a stretching circular tube of radius a moving at a velocity $w = 2cz$ in a stagnant free stream nanofluid. The physical properties of the fluid are assumed to be constant. Under such conditions, the governing equations of the steady, laminar boundary-layer flow on the moving surface are given by

$$\frac{\partial(rw)}{\partial z} + \frac{\partial(ru)}{\partial r} = 0, \tag{5.28}$$

$$w\frac{\partial w}{\partial z} + u\frac{\partial w}{\partial r} = \nu\left(\frac{\partial^2 w}{\partial r^2} + \frac{1}{r}\frac{\partial w}{\partial r}\right), \tag{5.29}$$

$$w\frac{\partial u}{\partial z} + u\frac{\partial u}{\partial r} = -\frac{1}{\rho_f}\frac{\partial p}{\partial r} + \nu\left(\frac{\partial^2 u}{\partial r^2} + \frac{1}{r}\frac{\partial u}{\partial r} - \frac{u}{r^2}\right), \tag{5.30}$$

$$w\frac{\partial T}{\partial z} + u\frac{\partial T}{\partial r} = \frac{\alpha}{r}\frac{\partial}{\partial r}\left(r\frac{\partial T}{\partial r}\right) + \tau\left\{D_B\left(\frac{\partial C}{\partial z}\frac{\partial T}{\partial z} + \frac{\partial C}{\partial r}\frac{\partial T}{\partial r}\right) + \left(\frac{D_T}{T_\infty}\right)\left[\left(\frac{\partial T}{\partial z}\right)^2 + \left(\frac{\partial T}{\partial r}\right)^2\right]\right\}, \tag{5.31}$$

$$w\frac{\partial C}{\partial z} + u\frac{\partial C}{\partial r} = \frac{D_B}{r}\frac{\partial}{\partial r}\left(r\frac{\partial C}{\partial r}\right) + \left(\frac{D_T}{T_\infty}\right)\frac{1}{r}\frac{\partial}{\partial r}\left(r\frac{\partial T}{\partial r}\right). \tag{5.32}$$

The boundary conditions are given by

$$r = a : u = 0, w = 2cz + \lambda\frac{\partial w}{\partial r}, T = T_w, C = C_w \tag{5.33a}$$

$$r \to \infty : w = 0, T = T_\infty, C = C_\infty. \tag{5.33b}$$

Proceeding with the analysis, we define the transformations

$$\eta = \left(\frac{r}{a}\right)^2, u = -ca\left(\frac{f(\eta)}{\sqrt{\eta}}\right), w = 2czf'(\eta), \theta = \frac{T - T_\infty}{T_w - T_\infty} \text{ and } \phi = \frac{C - C_\infty}{C_w - C_\infty}. \tag{5.34}$$

Using the transformation variables defined in equation (5.34), the governing transformed equations may be written as

$$\eta f''' + f'' + Re\left(ff''' - f'^2\right) = 0, \tag{5.35}$$

$$\frac{\theta''}{Pr} + \frac{\theta'}{\eta Pr} + Nb\theta'\phi' + Nt\theta'^2 + \frac{Re}{2\eta}f\theta' = 0, \tag{5.36}$$

$$\phi'' + \phi'\left(\frac{1}{2}LeRe f + \frac{1}{\eta}\right) + \frac{Nt}{Nb}\frac{1}{\eta}\theta' + \frac{Nt}{Nb}\theta'' = 0. \tag{5.37}$$

The transformed boundary conditions are given by

$$\eta = 1 : f = 0, f' = 1 + Bf'', \theta = 1, \phi = 1,$$
$$\eta \to \infty : f' = 0, \theta = 0, \phi = 0,$$

(5.38)

where $B = \dfrac{2\lambda}{a}$ is the slip parameter, prime denotes differentiation with respect to η, and the four parameters are defined by

$$Nt = \frac{\varepsilon(\rho c)_p D_T (T_w - T_\infty)}{(\rho c)_f T_\infty \alpha_m}, Nb = \frac{\varepsilon(\rho c)_p D_B (C_w - C_\infty)}{(\rho c)_f \alpha_m},$$

$$Re = \frac{ca^2}{2\nu}, Le = \frac{\nu}{D_B}, Pr = \frac{\nu}{\alpha}, \tau = \frac{\varepsilon(\rho c)_p}{(\rho c)_f}.$$

Here, Pr, Le, Nb, and Nt denote the Prandtl number, the Lewis number, the Brownian motion parameter, and the thermophoresis parameter, respectively. It is important to note that this boundary value problem reduces to the classical problem of flow, heat, and mass transfer due to a stretching cylinder in a viscous fluid when Nb and Nt are zero.

The quantities of practical interest in this study are the friction factor C_f, Nusselt number Nu, and the Sherwood number Sh.

The wall shear stress is given by

$$\tau_w = \mu \left(\frac{\partial w}{\partial r} \right)_{r=a} = \frac{4\mu cz}{a} \mu f''(1).$$

(5.39)

The friction factor C_f is given by

$$C_f = \frac{\tau_w}{\rho w^2 / 2} = 4 \frac{z}{a} \frac{1}{Re_z} f''(1)$$

(5.40)

where $Re_z = \dfrac{cz^2}{2\nu}$.

The local heat transfer rate (local Nusselt number) is given by

$$Nu_x = \frac{q_w x}{k(T_w - T_\infty)} = -2 \sqrt{\frac{Re_z}{Re}} \theta'(1).$$

(5.41)

Similarly the local Sherwood number is given by

$$Sh_x = \frac{q_m x}{D_B (C_w - C_\infty)} = -2 \sqrt{\frac{Re_z}{Re}} \phi'(1)$$

(5.42)

where q_w and q_m are wall heat and mass flux rates, respectively.

5.3.2 Numerical solutions

The nonlinear ordinary differential equations (5.35)–(5.37) satisfying the boundary conditions (5.38) were integrated numerically using the fourth-order Runge-Kutta scheme along with the shooting method for several values of the governing parameters. In order to assess the accuracy of the present results, we obtained results (in the absence of slip at the boundary) for the reduced friction factor $-f''(1)$ and Nusselt number $-\theta'(1)$ by ignoring the effects of Nb and Nt. These results are shown in Tables 5.3 and 5.4. A comparison of our results with literature values indicates excellent agreement, and therefore our results are highly accurate.

5.3.3 Discussion of the results

Figure 5.7(a)–(c) shows the velocity, temperature, and concentration distributions within the boundary layer as the Reynolds number Re increases. The momentum, thermal, and concentration boundary-layer thicknesses decrease as Re increases.

As Le increases, the temperature increases initially, but far away from the surface, temperature is found to decrease with increasing Le (Fig. 5.8(a)). It is observed that the concentration decreases and the concentration boundary-layer thickness decreases with increasing Le (Fig. 5.8(b)). The effect of Le is more pronounced for concentration fields than for temperature fields. This in turn increases the surface mass transfer rates as Le increases.

Table 5.3 **Comparison of results for $-f''(1)$**

Re_a	Present results	Wang [4]	Chamkha et al. [32]
0.5	0.88700	0.88220	0.88700
1.0	1.17923	1.17776	1.17953
2.0	1.59448	1.59390	1.59444
5.0	2.41755	2.41745	2.41798
10.0	3.34467		

Reprinted from Ref. [31].

Table 5.4 **Comparison of results for $-\theta'(1)$ ($Nt = Nb = 0$, $Re_a = 3$)**

Pr	Present results	Chamkha et al. [32]
0.70	1.15053	1.15053
2.00	2.10654	2.10655
7.00	4.23743	4.23743

Reprinted from Ref. [31].

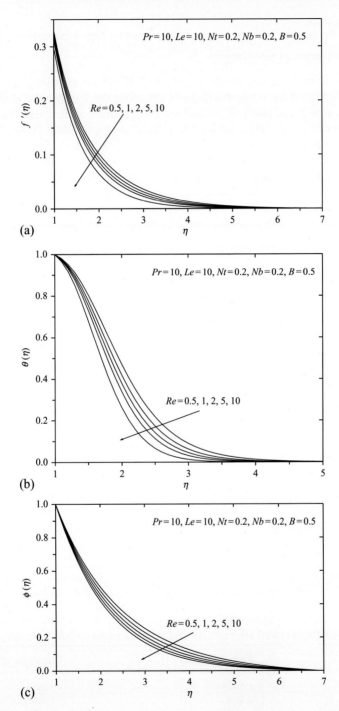

Figure 5.7 (a) Velocity, (b) temperature, and (c) concentration profiles for variable values of Reynolds number Re.
Reprinted from Ref. [31].

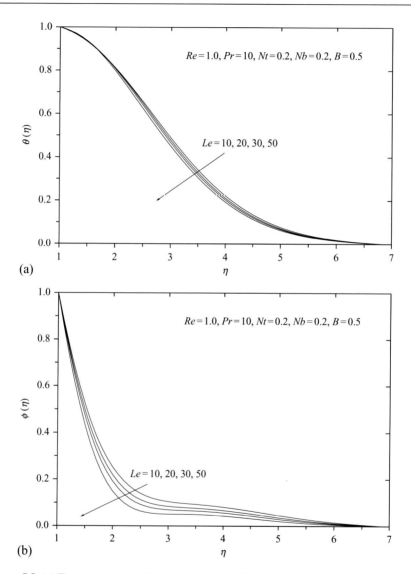

Figure 5.8 (a) Temperature and (b) concentration profiles for variable values of *Le*. Reprinted from Ref. [31].

References

[1] Pal D, Mondal H. Influence of temperature-dependent viscosity and thermal radiation on MHD forced convection over a non-isothermal wedge. Appl Math Comput 2009;212:194–208.
[2] Datta P, Anilkumar D, Roy S, Mahanti NC. Effect of non-uniform slot injection (suction) on a forced flow over a slender cylinder. Int J Heat Mass Tran 2006;49:2366–71.

[3] Kumari M, Nath G. Mixed convection boundary layer flow over a thin vertical cylinder with localized injection/suction and cooling/heating. Int J Heat Mass Transfer 2004;47:969–76.

[4] Wang CY. Fluid flow due to a stretching cylinder. Phys Fluids 1988;31:466–8.

[5] Ishak A, Nazar R, Pop I. Uniform suction/blowing effect on flow and heat transfer due to a stretching cylinder. Appl Math Model 2008;32:2059–66.

[6] Lin HT, Shih YP. Laminar boundary layer heat transfer along static and moving cylinders. J Chin Inst Eng 1980;3:73–9.

[7] Lin HT, Shih YP. Buoyancy effects on the laminar boundary layer heat transfer along vertically moving cylinders. J Chin Inst Eng 1981;4:47–51.

[8] Ishak A, Nazar R. Laminar boundary layer flow along a stretching cylinder. Eur J Sci Res 2009;36(1):22–9.

[9] Grubka LG, Bobba KM. Heat transfer characteristics of a continuous stretching surface with variable temperature. ASME J Heat Transfer 1985;107:248–50.

[10] Ali ME. Heat transfer characteristics of a continuous stretching. Surface, Heat Mass Transfer 1994;29:227–34.

[11] Ishak A, Nazar R, Pop I. Magnetohydrodynamic (MHD) flow and heat transfer due to a stretching cylinder. Energ Conv Manag 2008;49:3265–9.

[12] Bachok N, Ishak A. Flow and heat transfer over a stretching cylinder with prescribed surface heat flux. Malaysian J Math Sci 2010;4:159–69.

[13] Vajravelu K, Prasad KV, Santhi SR. Axisymmetric magneto-hydrodynamic (MHD) flow and heat transfer at a non-isothermal stretching cylinder. Appl Math Comput 2012;219:3993–4005.

[14] Mukhopadhyay S. MHD boundary layer slip flow along a stretching cylinder. Ain Shams Engng J 2013;4:317–24.

[15] Crane LJ. Flow past a stretching plate. Z Angew Math Phys 1970;21:645–7.

[16] Fang T, Zhang J, Yao S. Slip MHD viscous flow over a stretching sheet—an exact solution. Commun Nonlinear Sci Numer Simul 2009;14:3731–7.

[17] Wang CY. Flow due to a stretching boundary with partial slip—an exact solution of the Navier–Stokes equations. Chem Eng Sci 2002;57:3745–7.

[18] Andersson HI. Slip flow past a stretching surface. Acta Mech 2002;158:121–5.

[19] Anjali Devi SP, Ganga B. Effects of viscous and Joules dissipation on MHD flow, heat and mass transfer past a stretching porous surface embedded in a porous medium. Nonlinear Anal Model Control 2009;14:303–14.

[20] Mukhopadhyay S. Mixed convection boundary layer flow along a stretching cylinder in porous medium. J Petrol Sci Eng 2012;96–97:73–8.

[21] Mukhopadhyay S, Layek GC. Radiation effect on forced convective flow and heat transfer over a porous plate in a porous medium. Meccanica 2009;44:587–97.

[22] Mukhopadhyay S, De PR, Bhattacharyya K, Layek GC. Forced convective flow and heat transfer over a porous plate in a Darcy–Forchheimer porous medium in presence of radiation. Meccanica 2012;47:153–61.

[23] Ishak A, Nazar R, Pop I. Steady and unsteady boundary layers due to a stretching vertical sheet in a porous medium using Darcy–Brinkman equation model. Int J Appl Mech Eng 2006;11:623–37.

[24] Gorla RSR, Chamkha AJ, Rashad AM. Mixed convective boundary layer flow over a vertical wedge embedded in a porous medium saturated with a nanofluid: Natural Convection Dominated Regime. Nanoscale Res Lett 2011;6:207–18.

[25] Hamad MAA, Pop I. Scaling transformations for boundary layer flow near the stagnation-point on a heated permeable stretching surface in a porous medium saturated with a nanofluid and heat generation/absorption effects. Transp Porous Med 2011;87:25–39.

[26] Ding Y, Chen H, Wang L, Yang CY, He Y, Yang W, et al. Heat transfer intensification using nanofluids. KONA 2007;25:23–38.

[27] Eastman JA, Choi SUS, Li S, Yu W, Thompson LJ. Anomalously increased effective thermal conductivities containing copper nanoparticles. Appl Phys Lett 2001;78:718–20.

[28] Choi SUS, Zhang ZG, Yu W, Lockwood FE, Grulke EA. Anomalous thermal conductivity enhancement on nanotube suspension. Appl Phys Lett 2001;79:2252–4.

[29] Patel HE, Das SK, Sundararajan T, Sreekumaran A, George B, Pradeep T. Thermal conductivities of naked and monolayer protected metal nanoparticle based nanofluids: manifestation of anomalous enhancement and chemical effects. Appl Phys Lett 2003;83:2931–3.

[30] You SM, Kim JH, Kim KH. Effects of nanoparticles on critical Heat flux of water in pool boiling heat transfer. Appl Phys Lett 2003;83:3374–6.

[31] Mukhopadhyay S, Mondal IC, Gorla RSR. Effects of partial slip on boundary layer flow and heat transfer past a stretching circular cylinder in a nanofluid. Int J Microsc Nanosc Thermal Fluid Transport Phenom 2013;4:115–34.

[32] Chamkha AJ, Abd El-Aziz MM, Ahmed SE. Effects of thermal stratification on flow and heat transfer due to a stretching cylinder with uniform suction/injection. Int J Energy Technol 2010;2:1–7.

Flow past a sphere

6

A great breakthrough came in 1904 when Prandtl proposed his boundary-layer theory. By the term *boundary-layer flow*, we mean similar flows in the relatively thin regions near (next to) solid walls where the effect of viscosity is dominant; outside this layer, the flow may be regarded as that of an ideal fluid. Prandtl assumed that the fluid sticks to the wall of the body (no slip) where the flow velocity for a fixed body is zero, and the velocity increases from zero to the value of the free stream across a very thin layer. Thus, the wall-normal component of ∇u becomes large inside the boundary layer and the shear stress is relevant even for fluids of low viscosity. The shear stress at the wall exerts a drag force on the body known as the skin-friction drag. The boundary-layer theory thus offered an explanation for the conventional wisdom that a body moving through a fluid experiences a force. The boundary-layer theory had tremendous achievements to its credit and transformed fluid dynamics into a discipline of great engineering importance. When the external flow occurs at a curved surface, the development of the boundary layer is strongly affected by the corresponding normal pressure gradient toword the contour of curvature, giving rise to a crosswise flow. The physical explanation of this is as follows: In the free stream, the centrifugal force is counterbalanced by the pressure force. Toward the wall, the flow is retarded and the centrifugal force decreases, whereas the pressure is essentially unaffected. The resulting force induces a secondary flow in the cross-stream direction (the cross flow), which is maximum in the bulk of the boundary layer and vanishes at the edge of the layer.

6.1 Introduction and physical motivation

Convection heat and mass transfer in a porous medium from an axisymmetric body has attracted the attention of many investigators because of its wide range of applications in geophysics and energy-related problems, namely, thermal insulation, enhanced recovery of petroleum resource, geophysical flows, polymer processing in packed beds, and sensible heat storage bed. The design of a safe canister for nuclear waste disposal in the depth of the earth or in the sea bed demands a through understanding of the convective mechanism in porous media. In this direction, one needs to study the convective heat and mass transfer from different geometries. To begin with, axisymmetric bodies such as a cone, horizontal and vertical cylinder, and sphere are used to understand convective heat and mass transfer mechanisms.

Recently, the flow and heat transfer phenomena over a sphere have received considerable attention because of its practical applications in numerous problems. The analysis of heat transfer through a boundary-layer flow about a sphere has a wide range of technological applications, including solving the cooling problems in turbine blades, electronic systems, and manufacturing processes [1]. Chiang et al. [2] investigated the laminar free convection from a sphere by considering prescribed surface

Fluid Flow, Heat and Mass Transfer at Bodies of Different Shapes. http://dx.doi.org/10.1016/B978-0-12-803733-1.00006-5

temperature and surface heat flux. Dennis and Walker [3] studied the steady forced convection flow past a sphere at low and moderate Reynolds numbers. Chen and Mucoglu [4,5] extended the problem using the boundary-layer approximation with very large Reynolds and Grashof numbers. Free convection from a uniformly heated sphere, in which the fluid motion is generated as a result of buoyancy forces, has received much more attention. Potter and Riley [6] studied the free-convective flow from a heated sphere at large Grashof number. Convective heat transfer within fluid-saturated porous media has attracted considerable attention because of its ever-increasing applications in geophysics, oil recovery techniques, thermal insulation engineering, packed-bed catalytic reactors, and heat storage beds. A wide variety of these applications involving convective transport phenomena is cited by numerous authors. Cheng [7] investigated mixed convection flow about a cylinder and a sphere embedded in porous medium. Natural convection from a sphere with blowing and suction was studied by Huang and Chen [8]. El-Shaarawi et al. [9] investigated the mixed convection about a sphere between moderate and high Reynolds numbers with a wide range of viscosity ratios. Free convection at an axisymmetric stagnation point was considered by Amin and Riely [10]. Nazar et al. [11] extended the work of Chen and Mocuglo [4,5] at a constant surface temperature over the whole sphere, starting from the lower stagnation point up to the separation point. Hossain et al. [12] studied the conjugate effect of heat and mass transfer in natural convection flow from an isothermal sphere with chemical reaction. Molla et al. [13] investigated the effects of magnetohydrodynamic natural convection flow on a sphere in the presence of a heat source. In another paper, Molla et al. [14] extended the above problem by considering uniform heat flux. Later on, Mukhopadhyay [15] analyzed the effects of porous media near the lower stagnation point for flow past a sphere in the presence of a heat source/sink. We shall present here the results obtained by Mukhopadhyay [15].

6.2   Basic equations

We consider the two-dimensional steady natural convection flow of an incompressible viscous liquid over a sphere of radius a embedded in a porous medium.

The coordinates $\bar{x}$ and $\bar{y}$ measure respectively the distance along the surface of the sphere from the stagnation point and the distance normal to the surface of the sphere. A sketch of the physical problem and the coordinate system are shown in Fig. 6.1.

The governing boundary-layer equations with the application of Darcy's law and under the Boussinesq approximation in the usual notation are given by

$$\frac{\partial(\bar{r}\bar{u})}{\partial \bar{x}} + \frac{\partial(\bar{r}\bar{v})}{\partial \bar{y}} = 0, \tag{6.1}$$

$$\bar{u}\frac{\partial \bar{u}}{\partial \bar{x}} + \bar{v}\frac{\partial \bar{u}}{\partial \bar{y}} = g\beta(T - T_\infty)\sin\left(\frac{\bar{x}}{a}\right) + \nu\frac{\partial^2 \bar{u}}{\partial \bar{y}^2} - \frac{\nu}{k}\bar{u}. \tag{6.2}$$

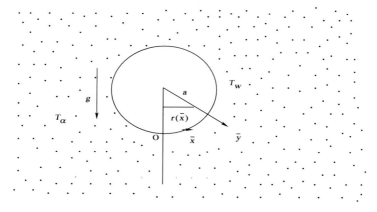

Figure 6.1 Sketch of the physical flow problem. Modified from Ref. [15].

Here $\bar{r}(\bar{x}) = a\sin\left(\dfrac{\bar{x}}{a}\right)$ is the radial distance from the symmetrical axis of the sphere, $\bar{u}$ and $\bar{v}$ are the components of velocity respectively in the $\bar{x}$ and $\bar{y}$ directions, k is the permeability of the porous medium, μ is the coefficient of fluid viscosity, ρ is the fluid density, $\nu = \frac{\mu}{\rho}$ is the kinematic viscosity, β is the coefficient of thermal expansion, and g is the acceleration due to gravity. By using the boundary-layer approximations and neglecting viscous dissipation, the energy equation is given by

$$\bar{u}\frac{\partial T}{\partial \bar{x}} + \bar{v}\frac{\partial T}{\partial \bar{y}} = \frac{\kappa}{\rho c_p}\frac{\partial^2 T}{\partial \bar{y}^2} + \frac{Q_0}{\rho c_p}(T - T_\infty), \qquad (6.3)$$

where T is the temperature, T_∞ is the free stream temperature, κ is the thermal conductivity, c_p is the specific heat at constant pressure, and Q_0 is the heat generation ($Q_0 > 0$) / absorption ($Q_0 < 0$) constant.

The appropriate boundary conditions for this problem are

$$\bar{u} = 0,\ \bar{v} = 0,\ T = T_w \ \text{ at } \ \bar{y} = 0, \qquad (6.4\text{a})$$

$$\bar{u} \to 0,\ T \to T_\infty \ \text{ as } \ \bar{y} \to \infty. \qquad (6.4\text{b})$$

Here T_w is the uniform surface temperature of the sphere. Constants T_w and T_∞ satisfy the inequality $T_w > T_\infty$.

6.3 Solution procedure

Let us introduce the dimensionless variables

$$x = \frac{\bar{x}}{a},\ y = (Gr)^{1/4}\frac{\bar{y}}{a},\ u = \frac{a}{\nu}(Gr)^{-1/2}\bar{u},\ v = \frac{a}{\nu}(Gr)^{-1/4}\bar{v},$$

$$\theta = \frac{T - T_\infty}{T_w - T_\infty},\ Gr = \frac{g\beta(T_w - T_\infty)a^3}{\nu^2},$$

where Gr is the Grashof number and θ is the nondimensional temperature.

The above equations (6.1)–(6.3) become

$$\frac{\partial(ru)}{\partial x} + \frac{\partial(rv)}{\partial y} = 0, \tag{6.5}$$

$$u\frac{\partial u}{\partial x} + v\frac{\partial u}{\partial y} = \theta \sin x + \frac{\partial^2 u}{\partial y^2} - \frac{a^2}{k(Gr)^{1/2}}u, \tag{6.6}$$

$$u\frac{\partial \theta}{\partial x} + v\frac{\partial \theta}{\partial y} = \frac{1}{Pr}\frac{\partial^2 \theta}{\partial y^2} + \frac{Q_0 a^2}{\nu \rho c_p (Gr)^{1/2}}\theta. \tag{6.7}$$

The boundary conditions take the form

$$u = 0, \ v = 0, \ \theta = 1 \ \text{ at } \ y = 0, \tag{6.8a}$$

$$u \to 0, \ \theta \to 0 \ \text{ as } \ y \to \infty. \tag{6.8b}$$

Introducing   nondimensional   stream   function $\psi = xr(x)f(x, y)$,   where
$u = \frac{1}{r}\frac{\partial \psi}{\partial y}, v = -\frac{1}{r}\frac{\partial \psi}{\partial x}.$

The equations (6.6-6.7) can be written as

$$\frac{\partial^3 f}{\partial y^3} + \left(1 + \frac{x\cos x}{\sin x}\right)f\frac{\partial^2 f}{\partial y^2} - \left(\frac{\partial f}{\partial y}\right)^2 + \theta\frac{\sin x}{x} - k_1\frac{\partial f}{\partial y}$$
$$= x\left(\frac{\partial f}{\partial y}\frac{\partial^2 f}{\partial y \partial x} - \frac{\partial f}{\partial x}\frac{\partial^2 f}{\partial y^2}\right), \tag{6.9}$$

$$\frac{1}{Pr}\frac{\partial^2 \theta}{\partial y^2} + \left(1 + \frac{x\cos x}{\sin x}\right)f\frac{\partial \theta}{\partial y} + \lambda\theta = x\left(\frac{\partial f}{\partial y}\frac{\partial \theta}{\partial x} - \frac{\partial f}{\partial x}\frac{\partial \theta}{\partial y}\right), \tag{6.10}$$

where $k_1 = \dfrac{a^2}{k(Gr)^{1/2}}$ is the permeability parameter of the porous medium,
$\lambda = \dfrac{Q_0 a^2}{\mu c_p (Gr)^{1/2}}$ is the heat source ($\lambda > 0$)/sink ($\lambda < 0$) parameter.

At the lower stagnation point, the above equations take the form

$$f''' + 2ff'' - f'^2 + \theta - k_1 f' = 0, \tag{6.11}$$

$$2f\theta' + \lambda\theta + \frac{1}{Pr}\theta'' = 0. \tag{6.12}$$

The boundary conditions become

$$f' = 0, f = 0, \theta = 1 \quad \text{at} \quad y = 0, \tag{6.13a}$$

$$f' \to 0, \theta \to 0 \quad \text{as} \quad y \to \infty. \tag{6.13b}$$

The above equations (6.11) and (6.12) along with the boundary conditions (6.13a, b) are solved by converting this boundary value problem to an initial value problem (Conte and Boor [16]).

6.4 Analysis of the result

Computation using the developed numerical scheme has been carried out for various values of the governing parameters, and the numerical values are plotted in Figs. 6.2 and 6.3. Our results (in the absence of porous media) are found to agree well with those of Molla et al. [13]. Horizontal velocity $f'(y)$ decreases with increase of the parameter of the porous medium (k_1) (Fig. 6.2(a)). The reason for this practical scenario is that the presence of porous medium causes higher restriction of the fluid flow which, in turn, slows its motion. But the temperature is found to increase with the increase of k_1 (Fig. 6.2(b)). The increase of permeability parameter k_1 leads to increase the skin-friction. The parameter of the porous medium k_1 introduces additional shear stress on the boundary, and as a result fluid velocity decreases, whereas the thermal boundary-layer thickness becomes thinner with the decreasing k_1, which implies that Darcian body force improves the heat transfer rate.

Figure 6.3(a) and (b) demonstrates the effects of heat source/sink on the velocity and temperature profiles. Fluid velocity increases with the increase of heat source/sink parameter λ (Fig. 6.3(a)). It is observed that, near the surface of the sphere, velocity increases significantly and then decreases slowly and finally vanishes asymptotically at $y = 6$. Temperature increases with the increase of the heat source parameter (Fig. 6.3 (b)). This feature prevails up to certain heights; then the process is slowed down and at a far distance from the surface of the sphere (at $y = 6$), temperature vanishes. On the other hand, the temperature field increases with the decrease in the amount of heat absorption. Again, far away from the surface of the sphere, such phenomena are smeared out (Fig. 6.3(b)). The presence of heat absorption ($\lambda < 0$) creates a layer of cold fluid adjacent to the heated surface of the sphere, and so the heat transfer rate from the surface of the sphere increases in this case.

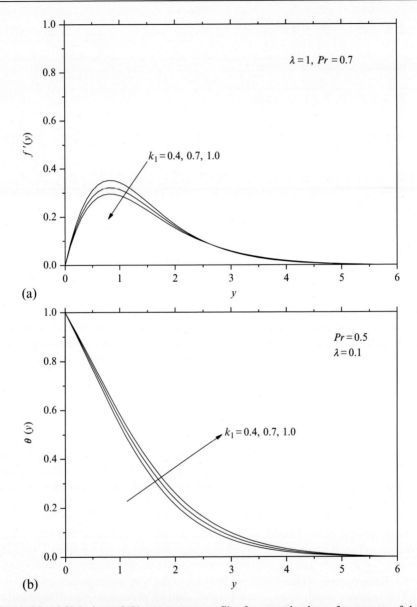

Figure 6.2 (a) Velocity and (b) temperature profiles for several values of parameter of the porous medium k_1. Modified from Ref. [15].

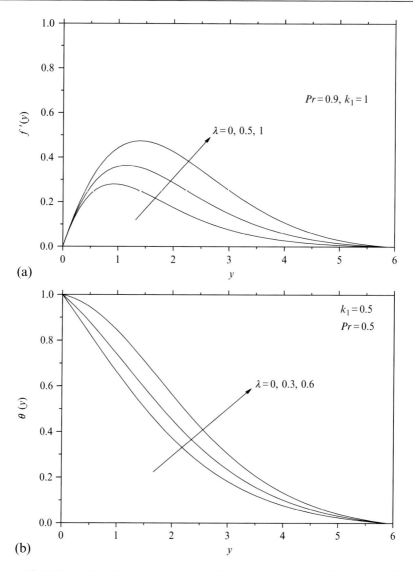

Figure 6.3 (a) Velocity and (b) temperature profiles for several values of heat source/sink parameter λ. Modified from Ref. [15].

6.5 Conclusions

Steady natural convection boundary-layer flow in the neighborhood of the lower stagnation point of a heated sphere embedded in a saturated porous medium in the presence of heat source/sink is considered. The dimensionless governing equations for this investigation are solved numerically by the shooting method (see, [15]).

It is our hope that the physics of flow over the surface of the sphere can be utilized as the basis for many scientific and engineering applications and for studying more complex problems of porous media.

References

[1] Nazar R, Amin N, Pop I. Mixed convection boundary layer flow about a sphere in a micropolar fluid with constant heat flux. J Energy Heat Mass Trans 2002;24:195–211.

[2] Chiang T, Ossin A, Tien CL. Laminar free convection from a sphere. ASME J Heat Tran 1964;86:537–42.

[3] Dennis SCR, Walker JDA. Calculation of the steady flow past a sphere at low and moderate Reynolds numbers. J Fluid Mech 1971;48:771–89.

[4] Chen TS, Mucoglu A. Analysis of mixed forced and free convection about a sphere. Int J Heat Mass Tran 1977;20:867–75.

[5] Chen TS, Mucoglu A. Mixed convection about a sphere with uniform surface heat flux. J Heat Trans 1978;100:542–4.

[6] Potter JM, Riley N. Free convection from a heated sphere at large Grashof number. J Fluid Mech 1980;100:769.

[7] Cheng P. Mixed convection about a horizontal cylinder and a sphere in a fluid saturated porous medium. Int J Heat Mass Transfer 1982;25:1245–7.

[8] Huang MJ, Chen CK. Laminar free convection from a sphere with blowing and suction. ASME J Heat Trans 1987;109:529–32.

[9] El-Shaarawi M, Ahmad NT, Kodah Z. Mixed convection about a rotating sphere in an axial stream. Numer Heat Trans A 1990;18:71–93.

[10] Amin N, Riley N. Free convection at an axisymmetric stagnation point. J Fluid Mech 1996;314:105–12.

[11] Nazar R, Amin N, Pop I. On the mixed convection boundary layer flow about a solid sphere with constant surface temperature. Arab J Sci Eng 2002;27:117–35.

[12] Hossain MA, Molla MM, Gorla RSR. Conjugate effect of heat and mass transfer in natural convection flow from an isothermal sphere with chemical reaction. Int J Fluid Mech Res 2004;31:104–17.

[13] Molla MM, Taher MA, Chowdhury MMK, Hossain MA. Magnetohydrodynamic natural convection flow on a sphere in presence of heat generation. Nonlinear Anal Model Control 2005;10:349–63.

[14] Molla MM, Hossain MA, Taher MA. Magnetohydrodynamic natural convection flow on a sphere with uniform heat flux in presence of heat generation. Acta Mechanica 2006;186:75–86.

[15] Mukhopadhyay S. Natural convection flow on a sphere through porous medium in presence of heat source/sink near a stagnation point. J Math Model Anal 2008;13:513–20.

[16] Conte SD, de Boor C. Elementary Numerical Analysis. New York: McGraw-Hill; 1972.

Flow past a wedge

7

During the last few decades, laminar boundary-layer flow past a wedge has been of considerable interest. Flow, heat and mass transfer along a wedge has gained considerable attention because of its vast applications in industry, and it has important bearings on several technological and natural processes. By employing the boundary-layer concept, several scientists obtained results that agree well with the experimental observations. Boundary-layer flows over wedge-shaped bodies are very common in many thermal engineering applications such as geothermal systems, crude oil extraction, ground water pollution, thermal insulation, heat exchangers, the storage of nuclear waste, etc. Within the field of aerodynamics, the analysis of boundary-layer problems for two-dimensional steady and incompressible laminar flow past a wedge is of particular interest.

7.1 Forced convection flow past a static wedge

The interest in self-similar solutions of the boundary-layer equation arose in the first half of the last century after Prandtl derived the boundary-layer equations in 1904. Falkner and Skan [1] derived the general equation for self-similar boundary layers and discussed the concept of the wedge phenomenon for two-dimensional flow. Later, Hartree [2] investigated the same problem and gave numerical results for wall shear stress for different values of the wedge angle. Stewartson [3] made an attempt to establish further solutions of the Falkner-Skan equation. Thereafter, many solutions have been obtained for different aspects of this class of boundary-layer problems [4–7]. A large amount of literature on this problem has been cited in the books by Schlichting and Gersten [8] as well as in Leal [9] and in the recent papers by Ishak et al. [10], Bararnia et al. [11], Parand et al. [12], Postelnicu and Pop [13], Afzal [14], and Ashwini and Eswara [15]. Koh and Hartnett [4] obtained the skin friction and heat transfer for incompressible laminar flow over a porous wedge with suction and variable wall temperature. Yih [16] presented an analysis of the forced convection boundary-layer flow over a wedge with uniform suction/blowing, whereas Watanabe [7] investigated the behavior of the boundary layer over a wedge with suction or injection in forced flow. We shall present here the results obtained by De et al. [17].

7.1.1 Mathematical analysis of the problem

We consider a steady, two-dimensional, laminar boundary-layer flow of viscous incompressible fluid past a symmetrical sharp porous wedge (Fig. 7.1) whose velocity is given by $\bar{u}_e(\bar{x}) = U_\infty \left(\frac{\bar{x}}{L}\right)^m$ for $m \leq 1$ where L is the characteristic length and m is wedge angle parameter related to the wedge angle $\pi\beta$ by $m = \frac{\beta}{2-\beta}$. For $m < 0$, the solution becomes singular at $\bar{x} = 0$, whereas for $m \geq 0$, the solution can be defined for all values of $\bar{x}$. The governing equations of such flow are, in the usual notation,

Fluid Flow, Heat and Mass Transfer at Bodies of Different Shapes. http://dx.doi.org/10.1016/B978-0-12-803733-1.00007-7

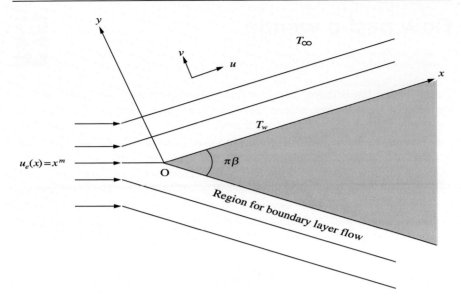

Figure 7.1 Sketch of the horizontal wedge with coordinate system [17].

$$\frac{\partial \bar{u}}{\partial \bar{x}} + \frac{\partial \bar{v}}{\partial \bar{y}} = 0, \tag{7.1}$$

$$\bar{u}\frac{\partial \bar{u}}{\partial \bar{x}} + \bar{v}\frac{\partial \bar{u}}{\partial \bar{y}} = \bar{u}_e\frac{\partial \bar{u}_e}{\partial \bar{x}} + \nu\frac{\partial^2 \bar{u}}{\partial \bar{y}^2}, \tag{7.2}$$

$$\bar{u}\frac{\partial T}{\partial \bar{x}} + \bar{v}\frac{\partial T}{\partial \bar{y}} = \kappa\frac{\partial^2 T}{\partial \bar{y}^2} + \frac{Q}{\rho C_p}(T - T_\infty), \tag{7.3}$$

where the viscous dissipation term in the energy equation is neglected because of its negligible effect in incompressible fluid motion. Here $\bar{u}$ and $\bar{v}$ are the components of velocity respectively in the $\bar{x}$ and $\bar{y}$ directions, $\nu = \frac{\mu}{\rho}$ is the kinematic viscosity of the fluid, μ is the coefficient of fluid viscosity, ρ is the fluid density, T is the temperature, κ is the coefficient of thermal diffusivity of the fluid, $Q = Q_0 x^{m-1}$ (Q_0 being a constant) is the volumetric rate of heat transfer, and c_p is the specific heat.

The appropriate boundary conditions for the problem are given by

$$u = 0, v = -\bar{v}_w, T = T_w \text{ at } \bar{y} = 0, \tag{7.4}$$

$$\bar{u} \to \bar{u}_e(\bar{x}), T \to T_\infty \text{ as } \bar{y} \to \infty, \tag{7.5}$$

where T_w is the wall temperature and T_∞ is the ambient temperature. Here the surface temperature T_w is higher than the ambient temperature T_∞ and $\bar{v}_w$ is the suction/ blowing velocity applied at the porous wall of the wedge.

7.1.2 Solution procedure

Introducing

$$x = \frac{\bar{x}}{L}, y = Re_L^{1/2}\frac{\bar{y}}{L}, u = \frac{\bar{u}}{U_\infty}, v = Re_L^{1/2}\frac{\bar{v}}{U_\infty},$$

$$u_\infty = \frac{\bar{u}_\infty}{U_\infty}, v_w = Re_L^{1/2}\frac{\bar{v}_w}{U_\infty}, Re_L = \frac{U_\infty L}{\nu} \tag{7.6}$$

in the above equations (7.1), (7.2), and (7.3) we get

$$\frac{\partial u}{\partial x} + \frac{\partial v}{\partial y} = 0, \tag{7.7}$$

$$u\frac{\partial u}{\partial x} + v\frac{\partial u}{\partial y} = u_e\frac{\partial u_e}{\partial x} + \frac{\partial^2 u}{\partial y^2}, \tag{7.8}$$

and

$$u\frac{\partial T}{\partial x} + v\frac{\partial T}{\partial y} = \frac{\kappa}{\nu}\frac{\partial^2 T}{\partial y^2} + \frac{QL}{U_\infty \rho c_p}(T - T_\infty). \tag{7.9}$$

The boundary conditions (7.4) and (7.5) now become

$$u = 0, v = -v_w, T = T_w \quad \text{at} \quad y = 0, \tag{7.10}$$

$$u \to u_e(x), T \to T_\infty \quad \text{as} \quad y \to \infty. \tag{7.11}$$

The velocity of the fluid over the wedge is now given by $u_e(x) = x^m$ for $m \le 1$. A special form of suction velocity $v_w = v_0 x^{\frac{m-1}{2}}$ (v_0 being a constant) is considered in this study.

We now introduce the relations

$$u = \frac{\partial \psi}{\partial y}, v = -\frac{\partial \psi}{\partial x} \text{ and } \theta = \frac{T - T_\infty}{T_w - T_\infty} \tag{7.12}$$

where ψ is the stream function.

Taking the relations (7.12) into consideration in the equations (7.8) and (7.9), we get

$$\frac{\partial \psi}{\partial y}\frac{\partial^2 \psi}{\partial x \partial y} - \frac{\partial \psi}{\partial x}\frac{\partial^2 \psi}{\partial y^2} = u_e\frac{\partial u_e}{\partial x} + \frac{\partial^3 \psi}{\partial y^3} = mx^{2m-1} + \frac{\partial^3 \psi}{\partial y^3}, \tag{7.13}$$

$$\frac{\partial \psi}{\partial y}\frac{\partial \theta}{\partial x} - \frac{\partial \psi}{\partial x}\frac{\partial \theta}{\partial y} = \frac{1}{Pr}\frac{\partial^2 \theta}{\partial y^2} + \lambda_1 \theta, \tag{7.14}$$

where $Pr = \dfrac{\nu}{\kappa}$ is the Prandtl number and $\lambda_1 = \dfrac{Q_0 x^{m-1} L}{\rho c_p U_\infty}$.

The boundary conditions (7.10) and (7.11) become

$$\frac{\partial \psi}{\partial y} = 0, \frac{\partial \psi}{\partial x} = v_w, \theta = 1 \quad \text{at} \quad y = 0 \tag{7.15}$$

and

$$\frac{\partial \psi}{\partial y} \to u_e(x) = x^m, \theta \to 0 \quad \text{as} \quad y \to \infty. \tag{7.16}$$

We now introduce the scaling group of transformations (Mukhopadhyay et al. [18]),

$$\Gamma : x^* = x e^{\varepsilon \alpha_1}, y^* = y e^{\varepsilon \alpha_2}, \psi^* = \psi e^{\varepsilon \alpha_3}, u^* = u e^{\varepsilon \alpha_4}, v^* = v e^{\varepsilon \alpha_5}, \theta^* = \theta e^{\varepsilon \alpha_6}. \tag{7.17}$$

Equation (7.16) may be considered as a point-transformation that transforms coordinates $(x, y, \psi, u, v, \theta)$ to the coordinates $(x^*, y^*, \psi^*, u^*, v^*, \theta^*)$.

Substituting (7.17) in (7.13) and (7.14) we get,

$$e^{\varepsilon(\alpha_1 + 2\alpha_2 - 2\alpha_3)} \left(\frac{\partial \psi^*}{\partial y^*} \frac{\partial^2 \psi^*}{\partial x^* \partial y^*} - \frac{\partial \psi^*}{\partial x^*} \frac{\partial^2 \psi^*}{\partial y^{*2}} \right) = m x^{*(2m-1)} e^{-\varepsilon(2m-1)\alpha_1} + e^{\varepsilon(3\alpha_2 - \alpha_3)} \frac{\partial^3 \psi^*}{\partial y^{*3}}, \tag{7.18}$$

$$e^{\varepsilon(\alpha_1 + \alpha_2 - \alpha_3 - \alpha_6)} \left(\frac{\partial \psi^*}{\partial y^*} \frac{\partial \theta^*}{\partial x^*} - \frac{\partial \psi^*}{\partial x^*} \frac{\partial \theta^*}{\partial y^*} \right) = \frac{1}{Pr} e^{\varepsilon(2\alpha_2 - \alpha_6)} \frac{\partial^2 \theta^*}{\partial y^{*2}} + e^{-\varepsilon \alpha_6} \lambda_1 \theta^*. \tag{7.19}$$

The system will remain invariant under the group of transformations Γ, so we have

$$\alpha_1 + 2\alpha_2 - 2\alpha_3 = -(2m-1)\alpha_1 = 3\alpha_2 - \alpha_3. \tag{7.20}$$

Solving (7.20) we get

$$\frac{\alpha_1}{\alpha_3} = \frac{2}{m+1}, \frac{\alpha_2}{\alpha_1} = \frac{1-m}{2}, \frac{\alpha_2}{\alpha_3} = \frac{1-m}{1+m}.$$

The boundary conditions become

$$\frac{\partial \psi^*}{\partial y^*} = 0, \frac{\partial \psi^*}{\partial x^*} = v_w e^{\varepsilon \frac{m-1}{2} \alpha_1}, \theta^* = 1 \quad \text{as} \quad y^* = 0, \tag{7.21}$$

$$\frac{\partial \psi^*}{\partial y^*} = x^{*m}, \theta^* \to 0 \quad \text{as} \quad y^* \to \infty \tag{7.22}$$

provided $\alpha_6 = 0$.

The set of transformations Γ now reduces to

$$x^* = xe^{\varepsilon\alpha_1}, y^* = ye^{\varepsilon\frac{1-m}{2}\alpha_1}, \psi^* = \psi e^{\varepsilon\frac{m+1}{2}\alpha_1}, \theta^* = \theta.$$

Expanding by Taylor's method in powers of ε and keeping terms up to the order ε, we get

$$x^* - x = x\varepsilon\alpha_1, y^* - y = y\varepsilon\frac{(1-m)\alpha_1}{2}, \psi^* - \psi = \psi\varepsilon\frac{(m+1)\alpha_1}{2}, \theta^* - \theta = 0.$$

In terms of differentials, these yield

$$\frac{dx}{\alpha_1 x} = \frac{dy}{\dfrac{(1-m)\alpha_1}{2}y} = \frac{d\psi}{\dfrac{(m+1)\alpha_1}{2}\psi} = \frac{d\theta}{0}.$$

Solving the above equations, we get

$$y^* x^{*\frac{(m-1)}{2}} = \eta, \psi^* = x^{*\frac{(m+1)}{2}}F(\eta), \theta^* = \theta(\eta). \tag{7.23}$$

With the help of equation (7.23), the equations (7.18) and (7.19) take the forms

$$mf'^2 - \frac{m+1}{2}ff'' - m = f''', \tag{7.24}$$

$$\theta'' + Pr\left[\frac{m+1}{2}f\theta' + \lambda\theta\right] = 0, \tag{7.25}$$

and the boundary conditions finally reduce to

$$f' = 0, f = S, \theta = 1 \quad \text{at} \quad \eta = 0 \tag{7.26}$$

and

$$f' = 1, \theta \to 0 \quad \text{as} \quad \eta \to \infty. \tag{7.27}$$

Here $\lambda = \frac{Q_0 L}{\rho c_p U_\infty}$ is the heat source/sink parameter and $S = \frac{2v_0}{m+1}$ is the suction $(S < 0)$ or blowing $(S < 0)$ parameter.

The above equations (7.24) and (7.25) along with the boundary conditions (7.26) and (7.27) are solved numerically by a shooting method.

We set

$$f' = z, z' = p, p' = \frac{2m}{m+1}(z^2 - 1) - fp, \tag{7.28}$$

Table 7.1 **Values of $f''(0)$ and $\theta'(0)$ for $S=0$, $\lambda = 0$ and $Pr=1$ [51]**

	$m=-0.05$		$m=0$		$m=\frac{1}{3}$	
	$f''(0)$	$\theta'(0)$	$f''(0)$	$\theta'(0)$	$f''(0)$	$\theta'(0)$
Present study	0.2135	0.2994	0.3320	0.3320	0.7574	0.4401
Pantokratoras [19] (with constant viscosity)	0.2135	0.2994	0.3320	0.3320	0.7574	0.4401

$$\theta' = q, q' = -Pr\left(\lambda\theta + \frac{m+1}{2}fq\right), \tag{7.29}$$

and the boundary conditions are $f'(0) = 0, f(0) = S, \theta(0) = 1$.

To integrate (7.28) and (7.29) as an initial value problem, we require $p(0)$, that is, $f''(0)$, and $q(0)$, that is, $\theta'(0)$, but no such values are given. The suitable guess values for $f''(0)$ and $\theta'(0)$ are chosen and the integration is carried out. We compare the calculated values for f' and θ at $\eta = 8$ (say) with the given boundary conditions $f'(8) = 1$ and $\theta(8) = 0$ and adjust the estimated values, $f''(0)$ and $\theta'(0)$, to give a better approximation for the solution.

We take a series of values for $f''(0)$ and $\theta'(0)$ and apply the fourth-order Runge-Kutta method with step size $h = 0.01$. To improve the solutions, we use linear interpolation, namely, the secant method. The above procedure is repeated until we get the results up to the desired degree of accuracy, 10^{-5}.

In order to assess the accuracy of the method, the results (in the absence of suction/blowing and heat source/sink) are compared with those of Pantokratoras [19]. Our results (in the absence of suction/blowing and heat source/sink, i.e., with $S=0$, $\lambda = 0$) are found to agree well with those of Pantokratoras [19] when the viscosity tends to a constant inside the boundary layer. For the sake of brevity, the comparison is made and the results are presented in Table 7.1.

7.1.3 Numerical results and discussion

In order to analyze the results, the numerical computation has been carried out using the method described in the previous section for various values of the parameters such as wedge angle parameter m, suction/blowing parameter S, and heat source/sink parameter λ. The results are plotted in Figs. 7.2–7.4.

Figure 7.2(a) presents the effects of wedge angle parameter (Falkner-Skan exponent) m on the fluid velocity when $S=0$. With increasing value of the exponent m, velocity is found to increase. It is found that in the case of accelerated flow ($m>0$), the velocity profiles have no point of inflection, whereas in the case of decelerated flow ($m<0$), they exhibit a point of inflection. Flow separation occurs at $m=-0.091$.

Figure 7.2(b) exhibits that the temperature $\theta(\eta)$ in boundary-layer region decreases with increasing values of the wedge angle parameter m ($S=0$, $\lambda=0$, and $Pr=0.2$).

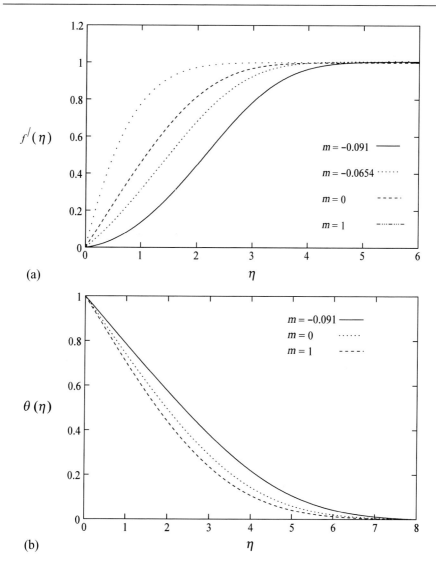

Figure 7.2 (a) Velocity and (b) temperature profiles for variable values of m when $S=0$ [17].

The thermal boundary-layer thickness decreases with m, resulting in an increase in the rate of heat transfer.

Figure 7.3(a) is the graphical representation of velocity distribution for various values of the suction/injection parameter S for $m = -0.091$, $\lambda = 0$, and $Pr = 0.2$. With the increasing S ($S > 0$), the horizontal velocity is found to increase. That is, suction results in increasing the velocity of the fluid, whereas due to injection ($S < 0$), fluid velocity decrease. Because the effect of suction is to suck away the fluid near the wall, the momentum boundary layer is reduced because of suction ($S > 0$). Consequently,

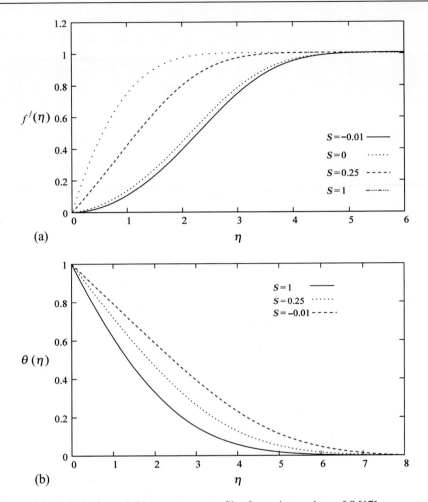

Figure 7.3 (a) Velocity and (b) temperature profiles for various values of S [17].

the velocity increases. Hence, the velocity gradient and so the skin friction increases with increasing S ($S>0$). The effect of injection ($S<0$) is opposite to that of suction ($S>0$). From this figure, it is very clear that flow separation can be controlled by the application of suction ($S>0$), whereas flow separation is influenced by injection ($S<0$). Therefore, by means of suction of the boundary layer through a porous wall, the velocity profile is made more stable. The stability of the velocity profile is favorable for the prevention of separation.

   Figure 7.3(b) exhibits the changes in the temperature field. This figure demonstrates that the temperature $\theta(\eta)$ decreases with increasing suction parameter S when $\lambda=0$ and $Pr=0.2$. The thermal boundary-layer thickness decreases with the suction parameter S, which causes an increase in the rate of heat transfer. The explanation for such a behavior is that the fluid is brought closer to the surface and it reduces the

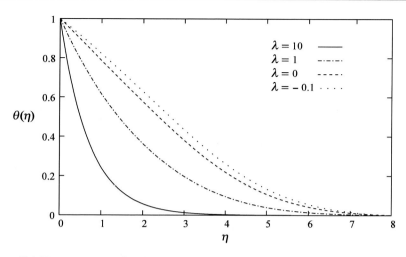

Figure 7.4 Temperature profiles for various values of heat source or sink parameter λ [17].

thermal boundary-layer thickness. But the temperature increases due to injection (for $S = -0.1$) with $\lambda = 0$ and $Pr = 0.2$. The thermal boundary-layer thickness increases with injection, which causes a decrease in the rate of heat transfer.

Figure 7.4 presents the influence of the heat generation (thermal source, $\lambda > 0$) or absorption (thermal sink, $\lambda < 0$) on the temperature profiles. Because the flow problem is uncoupled from the thermal problem, changes in the values of λ will not affect the fluid velocity. Physically, the presence of heat generation effects has the tendency to increase the fluid temperature, whereas in the case of heat absorption, the fluid temperature decreases as the thermal boundary layer becomes thinner.

7.2   Forced convection flow past a moving wedge

Most of the papers available in open literature are for the Falkner-Skan boundary-layer flow over a fixed wedge placed in a moving fluid. But boundary-layer behavior over a moving continuous solid surface is an important type of flow occurring in several engineering processes. In a very interesting paper, Riley and Weidman [20] and Ishak et al. [21] have studied multiple solutions of the Falkner-Skan equation for flow past a stretching boundary when the external velocity and the boundary velocity are each proportional to the same power law of the downstream distance. Recently, Khan and Pop [22] investigated the behavior of nanofluid flow over a moving wedge.

7.2.1   Formulation of the problem and solution procedure

Consider the steady two-dimensional mixed convective flow and heat transfer over a moving wedge in a moving viscous, incompressible fluid. The free stream velocity far away from the wedge is $u_e(x) = U_\infty \left(\frac{x}{L}\right)^m$ for $m \leq 1$ where L is a characteristic length and m is the wedge angle parameter related to the included angle $\pi\beta_1$ by $m = \frac{\beta_1}{(2-\beta_1)}$. The wedge is moving with a velocity $U_w(x) = Ru_e(x)$, where R is the velocity ratio parameter. The coordinate system is chosen such that the x-axis is along the wedge surface and the y-axis is normal to the wedge surface. The governing equations for such flow are

$$\frac{\partial u}{\partial x} + \frac{\partial v}{\partial y} = 0, \tag{7.30}$$

$$u\frac{\partial u}{\partial x} + v\frac{\partial u}{\partial y} = u_e\frac{\partial u_e}{\partial x} + \frac{\partial^2 u}{\partial y^2}, \tag{7.31}$$

and

$$u\frac{\partial T}{\partial x} + v\frac{\partial T}{\partial y} = \kappa\frac{\partial^2 T}{\partial y^2} \tag{7.32}$$

where $\nu = \frac{\mu}{\rho}$, μ is the coefficient of viscosity of the fluid, ρ is the fluid density, T is the temperature, and κ is the coefficient of thermal diffusivity of the fluid.

The boundary conditions now become

$$u = Ru_e(x), v = 0, T = T_w \quad \text{at} \quad y = 0, \tag{7.33}$$

$$u \to u_e(x), T \to T_\infty \quad \text{as} \quad y \to \infty. \tag{7.34}$$

The velocity of the fluid over the wedge is now given by $u_e(x) = x^m$ for $m \leq 1$. Here, T_w is the wall temperature and T_∞ is the ambient temperature.

We now introduce the relations

$$u = \frac{\partial \psi}{\partial y}, v = -\frac{\partial \psi}{\partial x}, \tag{7.35}$$

where ψ is the stream function.

We also introduce the variables

$$\theta(\eta) = \frac{T - T_\infty}{T_w - T_\infty}, \psi = \sqrt{\frac{2\nu U_\infty}{(m+1)L^m}} x^{\frac{m+1}{2}} f(\eta) \tag{7.36}$$

where

$$\eta = y\sqrt{\frac{(m+1)U_\infty}{2\nu L^m}} x^{\frac{m-1}{2}}. \tag{7.37}$$

With the help of the above relations, equation (7.30) is automatically satisfied, the equations (7.31) and (7.32) take the form

$$f''' + ff'' + \frac{2m}{m+1}\left(1 - f'^2\right) = 0,\tag{7.38}$$

$$\frac{1}{Pr}\theta'' + f\theta' = 0,\tag{7.39}$$

and the boundary conditions become

$$f'(\eta) = R, f(\eta) = 0, \theta(\eta) = 1 \text{ at } \eta = 0,\tag{7.40}$$

$$f'(\eta) - 1, \theta(\eta) \rightarrow 0 \text{ as } \eta \rightarrow \infty,\tag{7.41}$$

where $Pr = \frac{\nu}{\kappa}$ is the Prandtl number. The wedge moves faster than that of the free stream flow when $R > 1$ and moves slower than that of the free stream flow when $R < 1$. If $R < 0$, the wedge and fluid move in opposite direction whereas $R = 0$ presents the case of the static wedge.

The above equations (7.38) and (7.39) are solved numerically using the boundary conditions (7.40) and (7.41).

7.2.2 Results and discussion

To analyze the flow and heat transfer behaviors, the numerical data are computed for the governing parameters of the problem and are presented through Figs. 7.5(a)–7.6(b).

Figure 7.5(a)–(d) illustrates the effects of the velocity ratio parameter R on velocity profiles. An increase in R leads to a rise of velocity for $R < 1$ (Fig. 7.5(a)). Moreover, it displays that the fluid velocity is higher for higher values of wedge angle parameter m and the velocity boundary layer will thin when R approaches 1. Figure 7.5(b) exhibits the nature of velocity profiles for increasing R for $R > 1$. Here also fluid velocity increases with increasing values of R. But in this case fluid velocity is higher for lower values of the wedge angle parameter m (Fig. 7.5(b)).

Figure 7.5(c) and (d) display the flow behavior of the velocity field, respectively, for $m = 0.2$ and $m = -0.091$ when the wedge and fluid move in opposite directions, that is, when R takes negative values. Fluid velocity decreases with increasing magnitude of R ($R < 0$) for $m = 0.2$ (Fig. 7.5(c)). From Fig. 7.5(d), it is noted that though the velocity decreases initially with increasing magnitude of R, it finally increases with the increasing magnitude of R ($R < 0$). Velocity overshoot is noted for $R = -0.7, -0.9$. Peak value of velocity increases with the increasing magnitude of R (Fig. 7.5(d)).

Temperature and the thickness of the temperature boundary layer decrease with the increase of R (Fig. 7.6(a)). Temperature is found to increase with the increasing negative values of R (Fig 7.6(b)).

7.3 Mixed convection flow past a symmetric static/moving wedge

The mixed convection flow occurs in different industrial and technical applications, which include nuclear reactors cooled during emergency shutdown, electronic devices cooled by fans, solar central receivers exposed to wind currents, and heat exchangers placed in a low-velocity environment.

The effective heat transfer is required in a variety of energy technologies in order to enable the maximum possible power density and power conversion efficiency needed

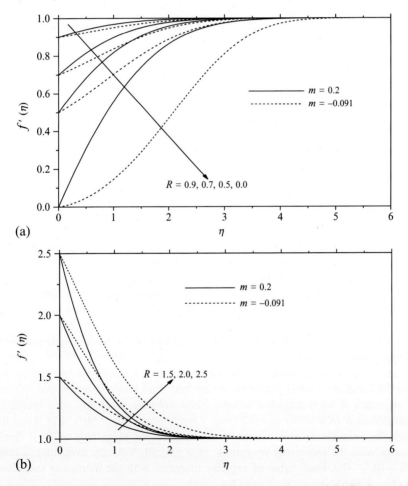

Figure 7.5 Velocity profiles for positive values of R when (a) $R < 1$ and (b) $R > 1$.

(Continued)

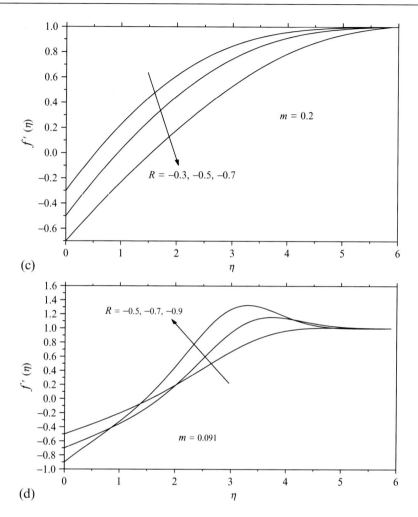

Figure 7.5 Continued. Velocity profiles for negative values of R when (c) $m=0.2$ and (d) $m=-0.091$.

for economic competitiveness and fuel conservation. Studies on response to boundary-layer flow and heat transfer along a symmetric wedge are of fundamental importance because of their vast applications in the industry and important bearings on several technological and natural processes. There are large numbers of investigations on free, forced, and mixed convective flow over a wedge (Smith [23], Williams and Rhyne [24], Hossain et al. [25]). Mixed convection in the laminar boundary-layer flow over a heated wedge or a plate is of considerable theoretical and practical interest. The number of studies on the mixed convective flow over a wedge is very limited. Sparrow et al. [26] were the first to investigate a combined forced and free convection flow and heat transfer about a nonisothermal wedge in a flow with a nonuniform

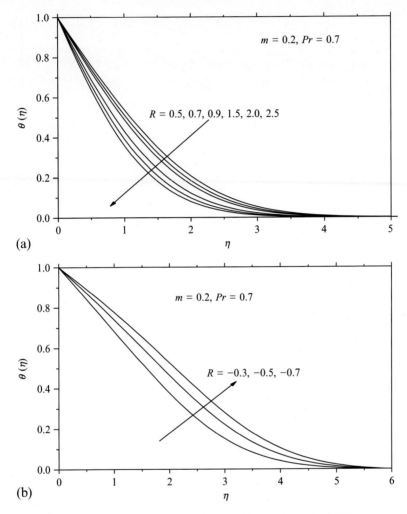

Figure 7.6 Temperature profiles for (a) positive and (b) negative values of R.

free-stream velocity. Bhattacharyya et al. [27] investigated unsteady mixed convection flow over a wedge embedded in porous medium. Nanousis [28] focused on the effect of suction or injection on a mixed convection flow past a wedge. Kumari et al. [29] investigated the mixed convection flow over a vertical wedge in a highly porous medium. Su et al. [30] investigated the effects of magnetic field on flow past a stretching wedge. Muhaimin et al. [31] investigated the unsteady MHD mixed convection flow past a porous wedge. Recently, Ganapathirao et al. [32] investigated the heat and mass transfer on unsteady mixed convection boundary-layer flow over a porous wedge with heat generation/absorption.

7.3.1 Mathematical formulation of the problem and solution procedure

Consider the steady mixed convection flow of an incompressible viscous fluid along a moving wedge. The wedge is moving with a velocity $U_w(x) = Ru_e(x)$, where R is the velocity ratio parameter. The free stream velocity far away from the wedge is $u_e(x) = U_\infty \left(\frac{x}{L}\right)^m$ for $m \leq 1$, where L is a characteristic length and m is the wedge angle parameter related to the included angle $\pi\beta_1$ by $m = \frac{\beta_1}{(2-\beta_1)}$.

The continuity, momentum, and energy equations governing such flow are

$$\frac{\partial u}{\partial x} + \frac{\partial v}{\partial y} = 0, \tag{7.42}$$

$$u\frac{\partial u}{\partial x} + v\frac{\partial u}{\partial y} = u_e\frac{\partial u_e}{\partial x} + v\frac{\partial^2 u}{\partial y^2} + g\beta(T - T_\infty), \tag{7.43}$$

$$u\frac{\partial T}{\partial x} + v\frac{\partial T}{\partial y} = \kappa\frac{\partial^2 T}{\partial y^2} \tag{7.44}$$

when the viscous dissipation term in the energy equation is neglected. Here u and v are the components of velocity respectively in the x and y directions, $v = \frac{\mu}{\rho}$ is kinematic viscosity, ρ is the fluid density, μ is the coefficient of fluid viscosity, β is the volumetric coefficient of thermal expansion, g is the gravity field, T is the temperature, T_∞ is the ambient temperature, and κ is the thermal diffusivity of the fluid.

The appropriate boundary conditions for the problem are given by

$$u = U_w(x) = Ru_e(x), v = 0, T = T_w(x) \quad \text{at} \quad y = 0, \tag{7.45}$$

$$u \to u_e(x), T \to T_\infty \quad \text{as} \quad y \to \infty. \tag{7.46}$$

Here the variable wall temperature $T_w(x)$ is given by $T_w(x) = T_\infty + T_0\left(\frac{x}{L}\right)^{2m-1}$, T_0 is a constant.

When $R > 1$, the wedge moves faster than that of the free stream flow, and it moves slower than that of the free stream flow when $R < 1$. If $R < 0$, the wedge and fluid move in opposite directions, whereas $R = 0$ presents the case of the static wedge.

Let us introduce

$$u = \frac{\partial \psi}{\partial y}, v = -\frac{\partial \psi}{\partial x}, \theta = \frac{T - T_\infty}{T_w - T_\infty}, \psi = \sqrt{\frac{2vU_\infty}{(m+1)L^m}}x^{\frac{m+1}{2}}f(\eta)$$

and

$$\eta = y\sqrt{\frac{(m+1)U_\infty}{2vL^m}}x^{\frac{m-1}{2}}, \tag{7.47}$$

where ψ is the stream function.

With the help of the above relations, equations (7.43) and (7.44) become

$$f''' + ff'' + \frac{2m}{m+1}\left(1 - f'^2\right) + \lambda\theta = 0, \tag{7.48}$$

$$\frac{1}{Pr}\theta'' + f\theta' - \frac{2(2m-1)}{(m+1)}f'\theta = 0, \tag{7.49}$$

where $\lambda = \frac{2g\beta T_0}{(m+1)U_\infty^2}$ is the mixed convection parameter, and $\lambda > 0$ aids the flow whereas $\lambda < 0$ opposes the flow.

The boundary conditions take the form

$$f'(\eta) = R, f(\eta) = 0, \theta(\eta) = 1 \text{ at } \eta = 0, \tag{7.50}$$

$$f'(\eta) = 1, \theta(\eta) \to 0 \text{ as } \eta \to \infty. \tag{7.51}$$

The above equations (7.48) and (7.49) are solved numerically with the help of the boundary conditions (7.50) and (7.51).

7.3.2   Analysis of results and concluding remarks

Numerical computations are made for variable values of the pertaining parameters and are plotted in Figs. 7.7(a)–7.8(c).

Fluid velocity increases as the mixed convection parameter λ increases, and velocity overshoot is noted for higher values of λ when the wedge and fluid move in the same direction (Fig. 7.7(a)). No overshoot is noted when they move in opposite directions. Fluid velocity at a particular point is higher when the fluid and wedge move in the same direction than that when they move in opposite directions. Temperature decreases with the increase in λ (Fig. 7.7(b)). But temperature at a point is higher when the fluid and wedge move in the opposite directions.

With increasing values of the velocity ratio parameter R, fluid velocity increases in both the cases of buoyancy-aided and -opposed flows (see Fig. 7.8(a)). Fluid velocity is higher for buoyancy-aided flow compared to that of buoyancy-opposed flow. But in both the cases, the temperature decreases with increasing values of R (Fig. 7.8(b) and (c)). Temperature is higher for buoyancy-opposed flow compared to buoyancy-aided flow (Fig. 7.8(b) and (c)). Thickness of the thermal boundary layer decreases with the increasing values of R.

7.4   Non-newtonian fluid flow over a symmetric wedge

During the last few decades, boundary-layer theory has been applied successfully in the case of non-Newtonian fluid. As the fluids are very complex in nature, no single constitutive equation exhibiting all properties of such fluids is available

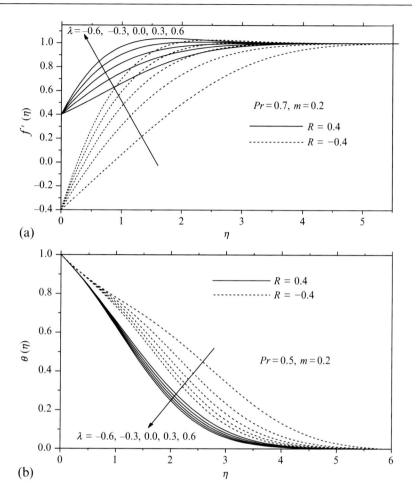

Figure 7.7 Effects of mixed convection parameter λ on (a) velocity and (b) temperature profiles.

(Mukhopadhyay [33], Mukhopadhyay and Vajravelu [34]). Several models can be found in this regard. As the governing equations of non-Newtonian fluids are highly nonlinear and much more complicated than that of Newtonian fluids, much care is needed in investigating such fluids for the understanding of the flow characteristics of a non-Newtonian fluid (Mukhopadhyay [35], Mukhopadhyay and Bhattacharyya [36]). The Casson fluid is a non-Newtonian fluid. In the literature, the Casson fluid model is sometimes stated to fit rheological data better than general viscoplastic models for many materials. It becomes the preferred rheological model for blood and chocolate. Casson fluid exhibits a yield stress. If a shear stress less than the yield stress is applied to the fluid, it behaves like a solid, whereas if a shear stress greater than the yield stress is applied, it starts to move (Eldabe and Elmohands [37], Dash et al. [38]).

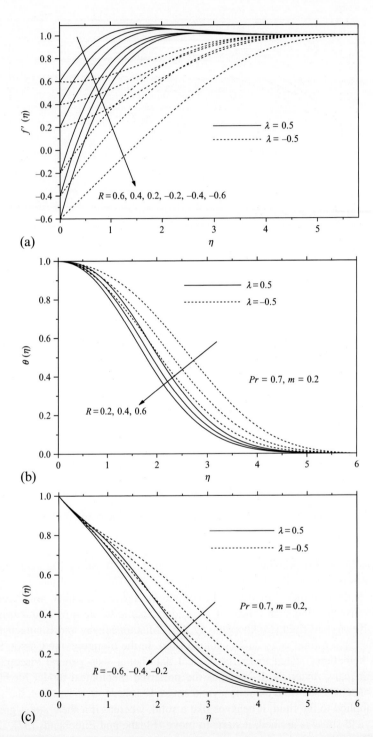

Figure 7.8 Effects of velocity ratio parameter R on (a) velocity and temperature when both wedge and fluid move in (b) same (c) opposite directions.

Keeping this in mind, many researchers investigated the non-Newtonian fluid flow past a wedge. Rajagopal et al. [39] studied the Falkner–Skan boundary-layer flow of a homogeneous incompressible second grade fluid past a wedge placed symmetrically with respect to the flow direction. Hady and Hassanien [40] reported the effect of a transverse magnetic field on a permeable symmetric wedge in a non-Newtonian fluid. Kafoussias and Nanousis [41] investigated the MHD laminar boundary-layer flow of a non-Newtonian fluid over a permeable wedge. Chamkha et al. [42] analyzed the free convective non-Newtonian fluid flow past a wedge embedded in a porous medium in the presence of thermal radiation. Later, Rashidi et al. [43] obtained the analytic solution for non-Newtonian flow and heat transfer over a nonisothermal wedge. Recently, Uddin et al. [44] investigated the micropolar fluid flow past a wedge in the presence of Hall and ion slip currents and analyzed the influences of thermal radiation and heat generation/absorption. Most of the works on the wedge flow problem in the literature are limited to the non-Newtonian power-law fluids. Recently, Mukhopadhyay et al. [45] discussed Casson fluid flow past a symmetric wedge. Here we shall present the results obtained by Mukhopadhyay and Mandal [46].

7.4.1 Formulation of the problem

We consider a steady, two-dimensional, laminar boundary-layer flow of a non-Newtonian Casson fluid past a symmetrical sharp wedge, and its velocity is given by $u_e(x) = U_\infty \left(\frac{x}{L}\right)^m$ for $m \leq 1$, where L is a characteristic length and m is the wedge angle parameter related to the included angle $\pi\beta_1$ by $m = \frac{\beta_1}{(2-\beta_1)}$. It is worth mentioning that β_1 is a measure of the pressure gradient. If β_1 is positive, the pressure gradient is negative or favorable, and a negative value of β_1 denotes a positive pressure gradient (adverse). For $m < 0$, the solution becomes singular at $x = 0$, whereas for $m \geq 0$, the solution can be defined for all values of x.

The rheological equation of state for an isotropic and incompressible flow of a Casson fluid is given by

$$\tau_{ij} = \begin{cases} 2\left(\mu_B + p_y/\sqrt{2\pi}\right)e_{ij}, & \pi > \pi_c \\ 2\left(\mu_B + p_y/\sqrt{2\pi_c}\right)e_{ij}, & \pi < \pi_c \end{cases}.$$

Here, $\pi = e_{ij}e_{ij}$ and e_{ij} is the (i,j)-th component of the deformation rate, π is the product of the component of deformation rate with itself, π_c is a critical value of this product based on the non-Newtonian model, μ_B is the plastic dynamic viscosity of the non-Newtonian fluid, and p_y is the yield stress of the fluid.

The governing equations of such flow, in the usual notation, are

$$\frac{\partial u}{\partial x} + \frac{\partial v}{\partial y} = 0, \tag{7.52}$$

$$u\frac{\partial u}{\partial x} + v\frac{\partial u}{\partial y} = u_e\frac{\partial u_e}{\partial x} + \nu\left(1 + \frac{1}{\beta}\right)\frac{\partial^2 u}{\partial y^2}, \tag{7.53}$$

$$u\frac{\partial T}{\partial x} + v\frac{\partial T}{\partial y} = \frac{\kappa}{\rho c_p}\frac{\partial^2 T}{\partial y^2}, \tag{7.54}$$

where the viscous dissipation term in the energy equation is neglected because of its small value for an incompressible fluid motion. Here, u and v are the components of velocity in the x and y directions, respectively. We recall that $\nu = \frac{\mu}{\rho}$ is the kinematic viscosity of the fluid, μ is the coefficient of fluid viscosity, ρ is the fluid density, β is the Casson parameter, U_∞ is the main stream velocity, T is the temperature, κ is the thermal conductivity of the fluid, and c_p is the specific heat at constant pressure.

The appropriate boundary conditions for the problem are given by

$$u = 0, v = -v_w(x), \frac{\partial T}{\partial y} = -\frac{q_w(x)}{\kappa} \quad \text{at} \quad y = 0 \tag{7.55}$$

$$u = u_e(x), T \to T_\infty \quad \text{as} \quad y \to \infty \tag{7.56}$$

where T_∞ is the free stream or ambient temperature. Here, $q_w(x) = \sqrt{\frac{2\nu U_\infty}{(m+1)L^m}}q_w x^{\frac{m-1}{2}}$ is the variable surface heat flux, $v_w(x) = v_0 x^{\frac{m-1}{2}}$, and a special type of velocity at the wall is considered where v_0 is a constant. Here $v_w(x) > 0$ is the velocity of suction and $v_w(x) < 0$ is the velocity of blowing.

We now introduce the relations

$$u = \frac{\partial \psi}{\partial y}, v = -\frac{\partial \psi}{\partial x} \tag{7.57}$$

where ψ is the stream function.

We also introduce the dimensionless variables

$$\theta(\eta) = \frac{\kappa(T - T_\infty)(m+1)}{q_w}\frac{(m+1)}{2\nu}, \text{ and } \psi = \sqrt{\frac{2\nu U_\infty}{(m+1)L^m}}x^{\frac{m+1}{2}}f(\eta) \tag{7.58}$$

where

$$\eta = y\sqrt{\frac{(m+1)U_\infty}{2\nu L^m}}x^{\frac{m-1}{2}}. \tag{7.59}$$

Using the relations (7.57) and (7.58) in the boundary-layer equation and in the energy equation, we get the equations

$$\frac{\partial \psi}{\partial y}\frac{\partial^2 \psi}{\partial x \partial y} - \frac{\partial \psi}{\partial x}\frac{\partial^2 \psi}{\partial y^2} = \frac{U_\infty^2}{L^{2m}}mx^{2m-1} + \nu\left(1 + \frac{1}{\beta}\right)\frac{\partial^3 \psi}{\partial y^3}, \tag{7.60}$$

$$\frac{\partial \psi}{\partial y}\frac{\partial \theta}{\partial x} - \frac{\partial \psi}{\partial x}\frac{\partial \theta}{\partial y} = \frac{\kappa}{\rho c_p}\frac{\partial^2 \theta}{\partial y^2}. \tag{7.61}$$

The boundary conditions then become

$$\frac{\partial \psi}{\partial y} = 0, \frac{\partial \psi}{\partial x} = v_w, \theta' = -1 \text{ at } y = 0, \tag{7.62}$$

$$\frac{\partial \psi}{\partial y} = \frac{U_\infty}{L^m} x^m, \theta \to 0 \text{ as } y \to \infty. \tag{7.63}$$

Using equation (7.59), equations (7.60) and (7.61) can be put in the form

$$\left(1 + \frac{1}{\beta}\right) f''' + ff'' + \frac{2m}{m+1} \left(1 - f'^2\right) = 0, \tag{7.64}$$

$$\frac{1}{Pr} \theta'' + f\theta' = 0, \tag{7.65}$$

and the boundary conditions become

$$f'(\eta) = 0, f(\eta) = S, \theta' = -1 \text{ at } \eta = 0, \tag{7.66}$$

$$f'(\eta) = 1, \theta \to 0 \text{ as } \eta \to \infty, \tag{7.67}$$

where $Pr = \frac{\mu c_p}{\kappa}$ is the Prandtl number, and $S = \frac{\sqrt{2}v_0}{\sqrt{\nu U_\infty (m+1)}}$ is the suction/blowing parameter.

Equations (7.64) and (7.65) along with the boundary conditions are solved by converting them to an initial value problem. We set

$$f' = z, z' = p, p' = \left\{\frac{2m}{m+1} \left(z^2 - 1\right) - fp\right\} \bigg/ \left(1 + \frac{1}{\beta}\right), \tag{7.68}$$

$$\theta' = q, q' = -Prfq, \tag{7.69}$$

with the boundary conditions

$$f(0) = S, f'(0) = 0, \theta'(0) = -1. \tag{7.70}$$

7.4.2 Numerical solutions

Equations (7.68) and (7.69) are solved numerically by the shooting technique. For the verification of the accuracy of the applied numerical scheme, results for $f''(0)$ for various values of m in case of a Newtonian fluid are compared with those reported by Yih [47], Chamkha et al. [48], Pal and Mondal [49], and Cebeci and Bradshaw [50]. The results are found to be in excellent agreement, which builds confidence that the present numerical results are accurate and the numerical method used is accurate. The comparisons are shown in Table 7.2.

Table 7.2 **Comparison of values of $f''(0)$ for variable values of m for Newtonian fluid [46]**

m	Yih [47]	Chamkha et al. [48]	Pal and Mondal [49]	Cebeci and Bradshaw [50]	Present study
−0.05	0.213484	0.213802	0.213484	0.21351	0.213802
0	0.332057	0.332206	0.332206	0.33206	0.332206
0.33	0.757448	0.757586	0.757586	0.75745	0.757586
1	1.232588	1.232710	1.232710	1.023259	1.232710

7.4.3　Discussion of the results

In order to analyze the results, numerical computation has been carried out for various values of the Falkner-Skan exponent m, Casson fluid parameter β, suction/blowing parameter S, and the Prandtl number Pr. For illustrations of the results, numerical values are plotted in Figs. 7.9(a) to 7.12.

Figure 7.9(a) and (b) depicts the effects of the Falkner-Skan exponent m on the velocity and temperature profiles in the presence of suction/blowing, respectively. Figure 7.9(a) presents the effects of increasing m on the fluid velocity. With increasing values of the exponent m, the fluid velocity is found to increase. It is observed that in case of an accelerated flow ($m > 0$), the velocity profiles have no point of inflection for both cases of suction and blowing. In the case of decelerated flow ($m < 0$), they exhibit a point of inflection that is noted in the presence of blowing ($S = -0.5$) only, and flow separation occurs at $m = -0.091$.

No point of inflection is noted in the case of suction ($S = 0.5$). For accelerated flows (i.e., positive values of m), the velocity profiles merely squeeze closer and closer to the wall, and backflow phenomena are not noted. Here, $m = 0$ presents the result for a flat plate. It is noted that the boundary-layer thickness decreases as m increases (Fig. 7.9(a)); hence, it gives rise to the velocity gradient, which in turn causes an increase in the skin friction. An increase in the wedge angle parameter leads to increase in the free stream velocity, and consequently the velocity boundary-layer thickness decreases. Figure 7.9(b) exhibits that the temperature $\theta(\eta)$ in the boundary-layer region decreases with increasing values of m. Temperature at a point is higher in case of blowing than that of suction. Temperature variation is more pronounced in the case of blowing. Because the thermal boundary-layer thickness increases with m, it causes a decrease in the rate of heat transfer (Fig. 7.9(b)).

The effects of the Casson fluid parameter β on the velocity and temperature profiles in the absence/presence of suction/blowing for accelerated flow are exhibited in Fig. 7.10(a) and (b), respectively. From the figures, it is very clear that with increasing values of β, the fluid velocity increases, but the temperature is found to decrease. Figure 7.10(a) clearly indicates that the thickness of the velocity boundary-layer decreases.

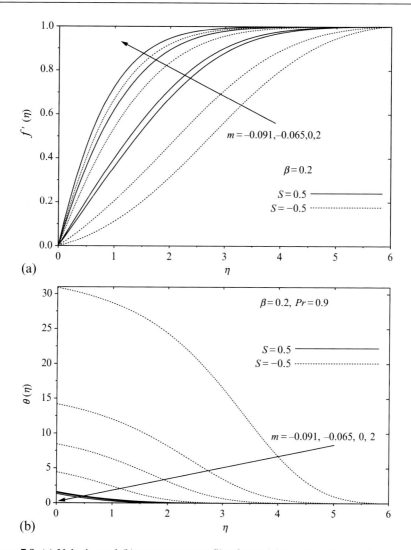

Figure 7.9 (a) Velocity and (b) temperature profiles for variable values of m in presence of suction/blowing [46].

Figure 7.11(a) presents the effects of the Casson parameter in the case of decelerated flow in the presence of suction/blowing. Fluid velocity is found to increase with increasing values of the Casson parameter β, in case of suction. For blowing, though the fluid velocity initially decreases with increasing values of β, it finally increases with increasing β (Fig. 7.11(a)). No point of inflection is observed for higher values of β in the presence of suction (see Fig. 7.11(a)). From this figure, it is very clear that

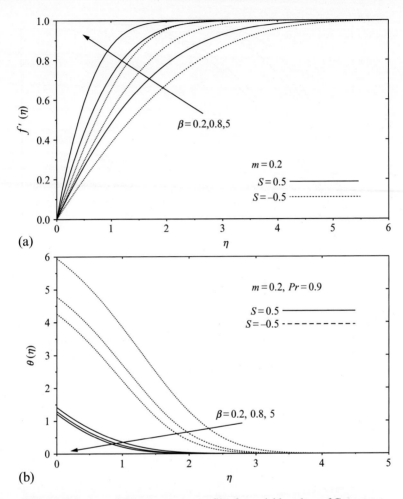

Figure 7.10 (a) Velocity and (b) temperature profiles for variable values of Casson parameter β for accelerated flow in presence of suction/blowing [46].

flow separation can be controlled in the case of a non-Newtonian Casson fluid by increasing the value of the Casson fluid parameter. Figure 7.11(b) show that the temperature decreases with increasing values of β. That is, the rate of heat transfer (the thermal boundary layer becomes thinner) is enhanced when the velocity boundary-layer thickness decreases.

Figure 7.12 exhibits the nature of the skin friction coefficient. It is very clear that the skin friction coefficient increases as the included angle of the wedge (β_1) increases; that is, the skin friction coefficient increases with the Falkner-Skan exponent m but it decreases with the Casson fluid parameter β.

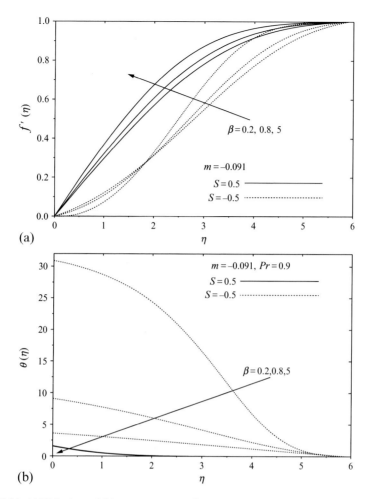

Figure 7.11 (a) Velocity and (b) temperature profiles for variable values of Casson parameter β in presence of suction/blowing for decelerated flow [46].

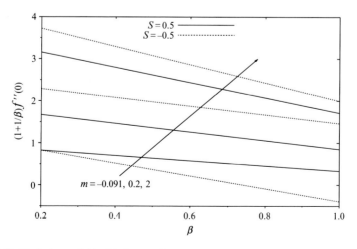

Figure 7.12 Variations of skin friction coefficient with Casson parameter β for various values of m in presence of suction/blowing [46].

References

[1] Falkner VM, Skan SW. Some approximate solutions of the boundary layer equations. Philos Mag 1931;12:865–96.

[2] Hartree DR. On an equation occurring in Falkner and Skan's approximate treatment of the equations of the boundary layer. Proc Camb Phil Soc 1937;33:223–39.

[3] Stewartson K. Further solutions of the Falkner-Skan equation. Proc Cambridge Phil Soc 1954;50:454–65.

[4] Koh JCY, Harnett JP. Skin friction and heat transfer for incompressible laminar flow over a porous wedge with suction and variable wall temperature. Int J Heat Mass Trans 1961;2:185–98.

[5] Chen KK, Libby PA. Boundary layers with small departure from the Falkner-Skan profile. J Fluid Mech 1968;33:273–82.

[6] Lin HT, Lin LK. Similarity solutions for laminar forced convection heat transfer from wedges to fluids of any Prandtl number. Int J Heat Mass Transfer 1987;30:1111–8.

[7] Watanabe T. Thermal boundary layer over a wedge with uniform suction and injection in forced flow. Acta Mech 1990;83:119–26.

[8] Schlichting H, Gersten K. Boundary layer theory. Eighth Revised Ed. Berlin: Springer-Verlag; 2000.

[9] Leal LG. Advanced transport phenomena: fluid mechanics and convective transport processes. New York: Cambridge University Press; 2007.

[10] Ishak A, Nazar R, Pop I. Falkner-Skan equation for flow past a moving wedge with suction or injection. J Appl Math Comput 2007;25:67–83.

[11] Bararnia H, Haghparast N, Miansari M, Barari A. Flow analysis for the Falkner–Skan wedge flow. Current Sci 2012;103:169–77.

[12] Parand K, Rezaei A, Ghaderi SM. An approximate solution of the MHD Falkner–Skan flow by Hermite functions pseudo spectral method. Commun Nonlinear Sci Numer Simul 2011;16:274–83.

[13] Postelnicu A, Pop I. Falkner–Skan boundary layer flow of a power-law fluid past a stretching wedge. Appl Math Comput 2011;217:4359–68.

[14] Afzal N. Falkner–Skan equation for flow past a stretching surface with suction or blowing, analytical solutions. Appl Math Comput 2010;217:2724–36.

[15] Ashwini G, Eswara AT. MHD Falkner-Skan boundary layer flow with internal heat generation or absorption. World Acad Sci Eng Tech 2012;65:662–5.

[16] Yih KA. Uniform suction/blowing effect on force convection about a wedge: uniform heat flux. Acta Mech 1998;128:173–81.

[17] De PR, Mukhopadhyay S, Layek GC. Analysis of fluid flow and heat transfer over a symmetric porous wedge. Acta Tech 2012;57:227–37.

[18] Mukhopadhyay S, Layek GC, Samad SA. Study of MHD boundary layer flow over a heated stretching sheet with variable viscosity. Int J Heat Mass Transfer 2005;48:4460–6.

[19] Pantokratoras A. The Falknar-Skan flow with constant wall temperature and variable viscosity. Int J Thermal Sci 2006;45:378–89.

[20] Riley N, Weidman PD. Multiple solutions of the Falkner-Skan equation for flow past a stretching boundary. SIAM J Appl Math 1989;49(5):1350–8.

[21] Ishak A, Nazar R, Pop I. Falkner-Skan equation for flow past a moving wedge with suction or injection. J Appl Math & Comput 2007;25:67–83.

[22] Khan WA, Pop I. Boundary layer flow past a wedge moving in a nanofluid. Math Prob Eng 2013, article ID 637285.

[23] Smith SH. The impulsive motion of a wedge in a viscous fluid. Z Angew Math Phys 1967;18:508–22.

[24] Williams JC, Rhyne TB. Boundary layer development on a wedge impulsively set into motion. SIAM J Appl Math 1980;38:215–24.

[25] Hossian MA, Bhowmik S, Gorla RSR. Unsteady mixed convection boundary layer flow along a symmetric wedge with variable surface temperature. Int J Eng Sci 2006; 44:607–20.

[26] Sparrow EM, Eichhorn R, Gregg JL. Combined forced and free convection in a boundary layer flow. Phys Fluids 1959;2:319–28.

[27] Bhattacharyya S, Pal A, Pop I. Unsteady mixed convection on a wedge in a porous medium. Int Commun Heat Mass Transfer 1998;25:743–52.

[28] Nanousis ND. Theoretical magnetohydrodynamic analysis of mixed convection boundary-layer flow over a wedge with uniform suction or injection. Acta Mech 1999;138:21–30.

[29] Kumari M, Takhar HS, Nath G. Mixed convection flow over a vertical wedge embedded in a highly porous medium. Heat Mass Transfer 2001;37:139–46.

[30] Su X, Zheng L, Zhang X, Zhang J. MHD mixed convective heat transfer over a permeable stretching wedge with thermal radiation and ohmic heating. Chem Eng Sci 2012;78:1–8.

[31] Muhaimin I, Kandasamy R, Khamis AB, Roslan R. Effect of thermophoresis particle deposition and chemical reaction on unsteady MHD mixed convective flow over a porous wedge in the presence of temperature-dependent viscosity. Nucl Eng Design 2013;261: 95–106.

[32] Ganapathirao M, Ravindran R, Momoniat E. Effects of chemical reaction, heat and mass transfer on an unsteady mixed convection boundary layer flow over a wedge with heat generation/absorption in the presence of suction or injection. Heat Mass Transfer 2014; http://dx.doi.org/10.1007/s00231-014-1414-1.

[33] Mukhopadhyay S. Upper-convected Maxwell fluid flow over an unsteady stretching surface embedded in porous medium subjected to suction/blowing. Z Naturforsch 2012;67a: 641–6.

[34] Mukhopadhyay S, Vajravelu K. Effects of transpiration and internal heat generation/ absorption on the unsteady flow of a Maxwell fluid at a stretching surface. ASME J Appl Mech 2012;79:044508-1–044508-6.

[35] Mukhopadhyay S. Heat transfer analysis for unsteady flow of a Maxwell fluid over a stretching surface in the presence of a heat source/sink. Chin Phys Lett 2012;29 (5):054703.

[36] Mukhopadhyay S, Bhattacharyya K. Unsteady flow of a Maxwell fluid over a stretching surface in presence of chemical reaction. J Egypt Math Soc 2012;20:229–34.

[37] Eldabe NTM, Salwa MGE. Heat transfer of MHD non-Newtonian Casson fluid flow between two rotating cylinders. J Phys Soc Jpn 1995;64:41–64.

[38] Dash RK, Mehta KN, Jayaraman G. Casson fluid flow in a pipe filled with a homogeneous porous medium. Int J Eng Sci 1996;34(10):1145–56.

[39] Rajagopal KR, Gupta AS, Na TY. A note on the Falkner-Skan flows of a non-Newtonian fluid. Int J Non-Linear Mech 1983;18:313–20.

[40] Hady FM, Hassanien IA. Effect of transverse magnetic field and porosity on the Falkner–Skan flows of a non-Newtonian fluid. Astrophys Space Sci 1985;112:381–90.

[41] Kafoussias NG, Nanousis ND. Magnetohydrodynamic laminar boundary-layer flow over a wedge with suction or injection. Can J Phys 1997;75:733–45.

[42] Chamkha AJ, Takhar HS, Beg OA. Radiative Free convective non-Newtonian fluid flow past a wedge embedded in a porous medium. Int J Fluid Mech Res 2004;31:101–15.

[43] Rashidi MM, Rastegari MT, Asadi M, B´eg OA. A study of non-Newtonian flow and heat transfer over a non-isothermal wedge using the homotopy analysis method. Chem Eng Commun 2012;231–56.

[44] Uddin Z, Kumar M, Harmand S. Influence of thermal radiation and heat generation// absorption on MHD heat transfer flow of a micropolar fluid past a wedge with hall and ion slip currents. Thermal Sci 2014;18:S489–502.

[45] Mukhopadhyay S, Mondal IC, Chamkha AJ. Casson fluid flow and heat transfer past a symmetric wedge. Heat Transfer Asian Res 2013;42:665–75.

[46] Mukhopadhyay S, Mandal IC. Boundary layer flow and heat transfer of a Casson fluid past a symmetric porous wedge with surface heat flux. Chin Phys B 2014;23(4):044702.

[47] Yih KA. Forced convection flow adjacent to a non-isothermal wedge. Int Commun Heat Mass Transfer 1996;26:819–27.

[48] Chamkha AJ, Quadri MM, Camille I. Thermal radiation effects on MHD forced convection flow adjacent to a non-isothermal wedge in the presence of a heat source or sink. Heat Mass Transfer 2003;39:305–12.

[49] Pal D, Mondal H. Influence of temperature-dependent viscosity and thermal radiation on MHD forced convection over a non-isothermal wedge. Appl Math Comput 2009;212: 194–208.

[50] Cebeci T, Bradshaw P. Physical and computational aspects of convective heat transfer. 1st ed. New York: Springer-Verlag; 1984.

[51] Mukhopadhyay S. Effects of radiation and variable fluid viscosity on flow and heat transfer along a symmetric wedge. J Appl Fluid Mech 2009;2(2):29–34.

Author Index

Note: Page numbers followed by *t* indicate tables.

Subject Index

Note: Page numbers followed by *f* indicate figures and *t* indicate tables.

Printed in the United States
By Bookmasters